1

Elements

SODIUM AND POTASSIUM

K

Na

Atlantic Europe Publishing

How to use this book

This book has been carefully developed to help you understand the chemistry of the elements. In it you will find a systematic and comprehensive coverage of the basic qualities of each element. Each two-page entry contains information at various levels of technical content and language, along with definitions of useful technical terms, as shown in the thumbnail diagram to the right. There is a comprehensive glossary of technical terms at the back of the book, along with an extensive index, key facts, an explanation of the Periodic Table, and a description of how to interpret chemical equations.

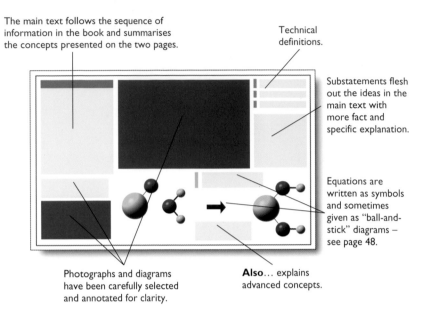

The main text follows the sequence of information in the book and summarises the concepts presented on the two pages.

Technical definitions.

Substatements flesh out the ideas in the main text with more fact and specific explanation.

Equations are written as symbols and sometimes given as "ball-and-stick" diagrams – see page 48.

Photographs and diagrams have been carefully selected and annotated for clarity.

Also... explains advanced concepts.

. . . .

An Atlantic Europe Publishing Book

Author
Brian Knapp, BSc, PhD
Project consultant
Keith B. Walshaw, MA, BSc, DPhil
(Head of Chemistry, Leighton Park School)
Industrial consultant
Jack Brettle, BSc, PhD (Chief Research Scientist, Pilkington plc)
Art Director
Duncan McCrae, BSc
Editor
Elizabeth Walker, BA
Special photography
Ian Gledhill
Illustrations
David Woodroffe
Electronic page make-up
Julie James Graphic Design
Designed and produced by
EARTHSCAPE EDITIONS
Print consultants
Landmark Production Consultants Ltd
Reproduced by
Leo Reprographics
Printed and bound by
Paramount Printing Company Ltd

Suggested cataloguing location
Knapp, Brian
Sodium and potassium
ISBN 1 869860 19 5
– *Elements* series
540

Acknowledgements
The publishers would like to thank the following for their kind help and advice: *Tim Fulford, Ian* and *Catherine Gledhill of Shutters, I. Made Rangun* and *Pippa Trounce.*

Picture credits
All photographs are from the **Earthscape Editions** photolibrary except the following:
(c=centre t=top b=bottom l=left r=right)
Tony Waltham, Geophotos 13b and **ZEFA** 14b

Front cover: A sodium light in a tunnel emits a characteristically orange colour.
Title page: A sample of rock salt, or halite, is principally composed of sodium and chlorine ions.

First published in 1996 by
Atlantic Europe Publishing Company Limited, Greys Court Farm,
Greys Court, Henley-on-Thames, Oxon, RG9 4PG, UK.

This product is manufactured from sustainable managed forests. For every tree cut down at least one more is planted.

The experiments described or illustrated in this book are not for replication. The Publisher cannot accept any responsibility for any accidents or injuries that may result from conducting the experiments described or illustrated in this book.

Contents

Introduction

An element is a substance that cannot be decomposed into a simpler substance by any known means. Each of the 92 naturally occurring elements is therefore one of the basic materials from which everything in the Universe is made. This book is about sodium and potassium.

Sodium

The white sparkling material that formed in salt pans of ancient Egypt was known as *natron*. We now call it soda, and from this word the element sodium takes its name. The chemical symbol for sodium is Na, after the word *natron*.

Sodium is the seventh most common element in the Universe and comprises one-thirtieth of the Earth's crust. On Earth its compounds give sea water its salty taste. The Egyptians used soda for embalming (much as we now use salt to preserve food). It was also used as a cleaning agent, and for making glass and pottery. Today, sodium vapour lamps give the yellow–orange lights used for street illumination. Sodium is also vital to regulating many body processes.

For all these reasons, sodium remains one of the most widely used elements.

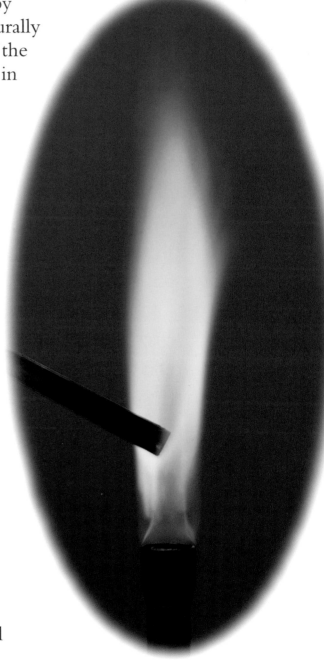

▲ This orange flame is produced when sodium is burned. This is the colour you will see in many street lights.

All in all, we consume about a hundred million tonnes of sodium compounds every year.

But although you can see sodium in many compounds, it is rare to see the soft, silvery-coloured metal that is pure sodium because it reacts violently in air and water. In nature, sodium is always found combined with other elements.

Potassium

Potassium, the twentieth most common element in the Universe and the seventh most common on Earth, is a lightweight, soft, silvery-coloured metal, whose name comes from the word *potash*. The chemical symbol, K, however, comes from the Latin word *kalium*, which in turn comes from the Arabic word for alkali.

Potassium and sodium have many similar properties and occur in the same group in the pattern of elements called the Periodic Table (see page 46). Like sodium, potassium is too reactive to occur naturally and is always found in compounds. It reacts even more violently than sodium when exposed to air and water, meaning that demonstrations with potassium can be dangerous.

The most common uses of potassium are as potash the compound used in glass-making and as a fertiliser.

▲ This lilac-coloured flame is produced when potassium is burned. It is one way to distinguish the presence of the element potassium.

Sodium

Sodium is a soft, silvery-coloured metal that can be obtained by passing an electric current through molten table salt (sodium chloride).

Sodium does not occur in nature as a metal because it is so reactive. For example, sodium will react violently with water, as this demonstration shows.

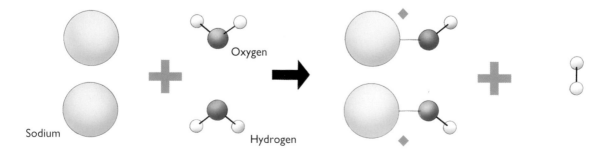

Sodium pellet

❶ A silvery pellet of sodium. Sodium corrodes quickly in air and so has few uses.

❷ The sodium pellet is dropped onto water in a small dish. A chemical indicator has been added to the water, partly to show the trail of the sodium as it moves across the water, and also to show what happens as the sodium reacts. The indicator is colourless unless the solution becomes alkaline, in which case it turns bright pink.

Also...

The reaction between water and sodium metal produces sodium hydroxide and releases hydrogen gas (the fizz). Sodium hydroxide, also known as caustic soda (see page 32), is a strongly alkaline compound. The alkaline reaction causes the phenolphthalein indicator to turn pink.

EQUATION: Reaction of sodium metal in water

Sodium + water ⇨ sodium hydroxide + hydrogen

$2Na(s) + 2H_2O(l) \Rightarrow 2NaOH(aq) + H_2(g)$

Sodium

Oxygen

Hydrogen

alkaline: the opposite of acidic. Alkalis are bases that dissolve, and alkaline materials are called basic materials. Solutions of alkalis have a pH greater than 7.0 because they contain relatively few hydrogen ions.

corrosion: the *slow* decay of a substance resulting from contact with gases and liquids in the environment. The term is often applied to metals. Rust is the corrosion of iron.

indicator: a substance or mixture of substances that change colour with acidity or alkalinity.

reactivity: the tendency of a substance to react with other substances. The term is most widely used in comparing the reactivity of metals. Metals are arranged in a reactivity series.

❸ The pellet immediately begins to fizz (react) on the surface, skimming about and creating a trail that can be seen in this enlarged picture. The heat released causes the metal to form a rolling molten ball. Notice that the water has turned bright pink.

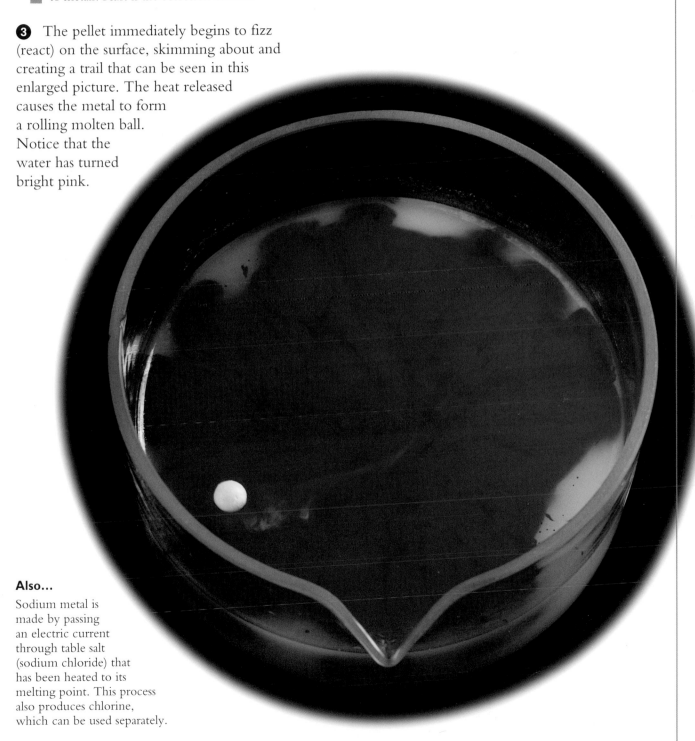

Also...

Sodium metal is made by passing an electric current through table salt (sodium chloride) that has been heated to its melting point. This process also produces chlorine, which can be used separately.

Crystals containing sodium

All the sodium compounds that are now on the surface of the Earth originated in volcanic rocks such as granite. Sodium is present in the white and pink feldspar crystals in a piece of granite. When volcanic rocks decompose they release sodium.

Sodium compounds are found in all the freshwater rivers and lakes of the world in tiny amounts. But when the lake dries out, the sodium compounds are left behind. The most common is the white sparkling mineral (sodium chloride) that we know as common salt, or rock salt.

Crystals of salt are soft and easily scratched. They are mostly white, although they may be tinged with orange if they have some iron staining. The crystals are cubic (box-shaped), and are made up of sodium combined with another element, chlorine.

Minerals dissolved in water are left behind when it evaporates, leaving salt deposits. These contain a variety of compounds, but they are usually dominated by sodium chloride. Salt deposits are common in most deserts and also in salt marshes near estuaries. Geologists refer to rock salt as "halite".

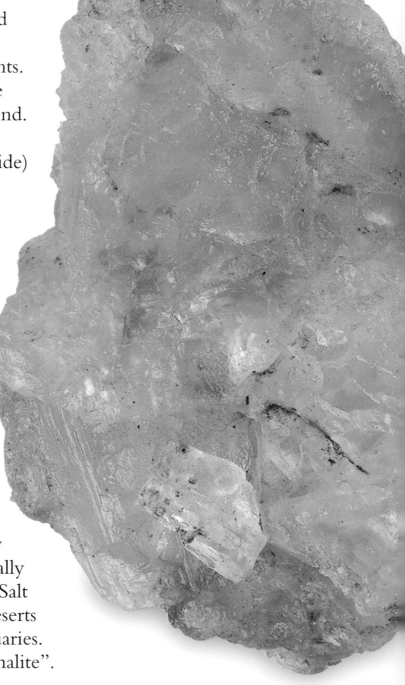

► The best place to look for crystals of sodium chloride is in the table salt kept in every home. The crystals can be seen with a magnifying glass. The best place to look for crystals outside the home is in those places where saline solutions evaporate, such as coastal lagoons or desert lakes. The picture on the right is of the surface of one of the world's best-known salt deposits, on the floor of Death Valley, California (see also page 11). Notice how the crystals have grown up into tall square-sectioned protrusions. Thus they retain their cubic structure.

crystal: a substance that has grown freely so that it can develop external faces. Compare with crystalline, where the atoms are not free to form individual crystals and amorphous where the atoms are arranged irregularly.

crystalline: the organisation of atoms into a rigid "honeycomb-like" pattern without distinct crystal faces.

decompose: to break down a substance (for example by heat or with the aid of a catalyst) into simpler components. In such a chemical reaction only one substance is involved.

mineral: a solid substance made of just one element or chemical compound. Calcite is a mineral because it consists only of calcium carbonate, halite is a mineral because it contains only sodium chloride.

translucent: almost transparent.

vitreous: glass-like.

◄ The most common compound of sodium is rock salt or halite. Rock salt is a glassy-looking (vitreous) material.
Shown here is a crystalline mass of rock salt taken from a salt mine. It is not easy to see individual crystals in crystalline material.

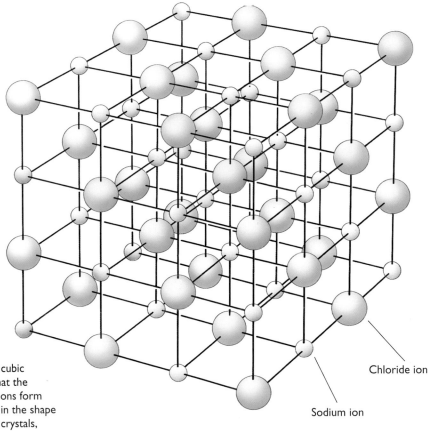

Chloride ion

Sodium ion

► As crystals of salt grow, they form cubic crystals. In this diagram you can see that the sodium (yellow) and chloride (green) ions form a regular framework. This is reflected in the shape of individual crystals. Thus, even large crystals, made of millions of sodium and chloride ions, have a cubic crystal structure.

Salt lakes

Geologists call dry desert lake beds "playas" after the Spanish word for beach. It is likely that such lakes produced thick beds of salt in the past, too. These have since been buried by other rocks and the minerals have become compacted into rock. One layer of such minerals makes up rock salt, the mineral described on the previous page.

▲ This picture shows a playa lake in Death Valley.

Salt shores

The white band along this lake shore is salt. It has been produced by the heat of the sun on the sediments surrounding the lake. As the sun heats the sediment, it draws water from the lake and up to the surface, where it is evaporated. The minerals in the solution are left behind as a white crust, most of which is salt. Salty shorelines by the sea are produced by the same effect.

▲ The picture above shows a shoreline where evaporation is intense. The white deposits are salt.

Salt lakes

Salt lakes occur throughout the world's desert regions. One of the largest shallow lakes in the United States is the Great Salt Lake. It has an area of about 5000 sq km. This lake receives water from nearby rivers, but water evaporates as fast as it enters. Nearby are playas such as the Bonneville salt flats.

Some of the lakes of Africa's Great Rift Valley are also salt lakes. Even larger playas exist in Australia. Lake Eyre, in central South Australia, covers 9000 sq km and is the lowest point in Australia. Its salt crust is about 15 cm thick.

evaporation: the change of state of a liquid to a gas. Evaporation happens below the boiling point and is used as a method of separating out the materials in a solution.

As the saline water in the playa evaporates, delicate salt crystals form as a crust.

▼ The white deposits shown in the picture below are salt. They are toxic to most plants.

Mining salt

Salt is one of the most important minerals used today. It is used as pure sodium chloride and also as the raw material for obtaining sodium and chlorine for other processes.

Salt occurs in thick rock beds and also in sea water. Salt pans (evaporating sea salt) produce only a small amount of the world's need for salt. This is because salt is used in greater quantities and for more applications than any other mineral. Most salt is therefore obtained from underground rocks.

Solution mining

Some salt beds cannot be mined by blasting. Instead the salt is recovered by dissolving it. First a deep well is sunk down to the salt bed or dome. Hot water is then fed down the outside of the well. The water dissolves the salt, and the resulting brine is drawn back to the surface.

When the brine reaches the surface, the water is evaporated away using special vacuum flasks.

▶ A diagrammatic representation of how salt is brought to the surface as a brine solution. The salt may occur as beds or as salt domes.

Deep mining

Most salt is obtained from salt beds deep underground. In some cases it is possible to mine the salt in a fashion something like coal mining. A shaft is sunk and horizontal galleries are opened out into the salt beds. The salt is a true rock and is recovered by blasting it and then hoisting it to the surface.

Once the rock salt has reached the surface it is crushed to different sizes, depending on its intended use.

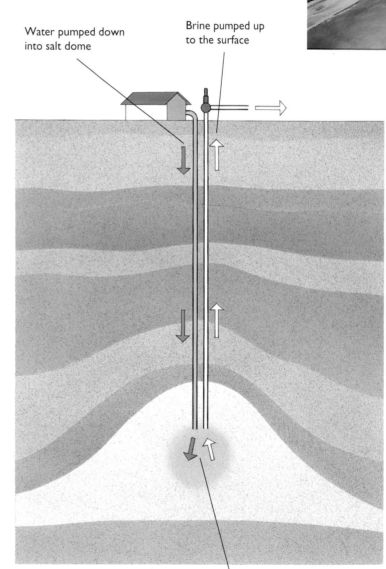

Water pumped down into salt dome

Brine pumped up to the surface

Water dissolves rock salt in a dome to make a concentrated salt solution called brine. This can then be drawn back up to the surface.

▲ This picture shows coastal salt pans. The salt has been collected and formed into the white central pile. From there it is moved away by truck and train to customers nationwide.

► So-called "sea-salt", obtained from evaporation, can be produced in large crystals. They are then ground down for use on the table.

Salt from the sea

Common salt makes up four-fifths of the minerals dissolved in sea water.

Salt has been collected since ancient times. To do this, people have used a method that copies the playas. They make shallow salt pans next to estuaries, each with a mud wall. Sea water is then allowed into the salt pans. The water evaporates under the heat of the sun, becoming very salty "brine".

Gradually, salt crystals form and settle out at the bottom of the brine. The rest of the water, known as the bittern, is then drawn off and the salt is allowed to dry off before it is scraped up and carted away.

About 200 metres of sea water have to evaporate in order to get one metre of salt, so getting salt from salt pans is a slow process.

The crystals of salt that grow in salt pans are much larger than salt produced in any other way. This is why sea-salt crystals have to be crushed in a salt grinder.

► This picture shows an underground rock salt mine.

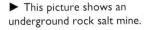

Salt and osmosis

Osmosis is one of the most important processes in the natural world. It describes the way that a liquid can pass through some materials but substances dissolved in it may be held back. Osmosis plays a vital part in industrial chemistry and is used to act as a natural "sieve". This kind of sieve is known by scientists as a "semipermeable membrane". A common example of a semipermeable membrane is human skin. The tissue walls of many of our internal organs are also semipermeable.

If a concentrated solution and a dilute solution are put together, they will mix until the whole solution is the same concentration. (For example, a dye will spread through a liquid). The semipermeable membrane works because the molecules of the liquid part of a solution are small, but the molecules of the substances dissolved in the liquid are large. The membrane has holes in it just large enough to allow the liquid molecules to pass through easily, but just too small to allow the molecules of the dissolved substances.

The principles involved in osmosis are demonstrated here using salt.

Desalination

A process called reverse osmosis is used to remove salts from impure natural water. This is how desalination plants work.

Salty water is pumped through tubes made of a semipermeable membrane (often a kind of synthetic rubber). Pure water is collected from outside the membrane.

▼ This is a desalination plant in the Middle East, a region naturally very short of water but with enough oil reserves to use an energy-intensive method of collecting fresh water.

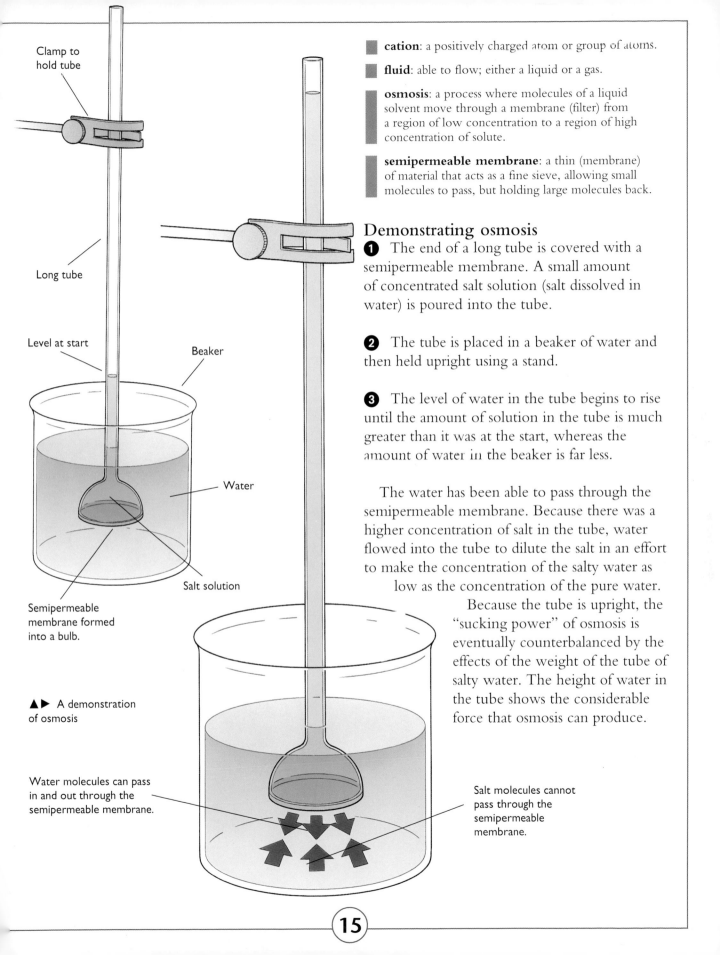

Clamp to
hold tube

Long tube

Level at start

Beaker

Water

Salt solution

Semipermeable
membrane formed
into a bulb.

▲▶ A demonstration
of osmosis

Water molecules can pass
in and out through the
semipermeable membrane.

Salt molecules cannot
pass through the
semipermeable
membrane.

cation: a positively charged atom or group of atoms.

fluid: able to flow; either a liquid or a gas.

osmosis: a process where molecules of a liquid solvent move through a membrane (filter) from a region of low concentration to a region of high concentration of solute.

semipermeable membrane: a thin (membrane) of material that acts as a fine sieve, allowing small molecules to pass, but holding large molecules back.

Demonstrating osmosis

❶   The end of a long tube is covered with a semipermeable membrane. A small amount of concentrated salt solution (salt dissolved in water) is poured into the tube.

❷   The tube is placed in a beaker of water and then held upright using a stand.

❸   The level of water in the tube begins to rise until the amount of solution in the tube is much greater than it was at the start, whereas the amount of water in the beaker is far less.

The water has been able to pass through the semipermeable membrane. Because there was a higher concentration of salt in the tube, water flowed into the tube to dilute the salt in an effort to make the concentration of the salty water as low as the concentration of the pure water.

Because the tube is upright, the "sucking power" of osmosis is eventually counterbalanced by the effects of the weight of the tube of salty water. The height of water in the tube shows the considerable force that osmosis can produce.

Salt and living things

In living things *all* cells are surrounded by semipermeable membranes. The walls of the cells are able to filter out some molecules and let others pass through. In this way they can eject waste materials and absorb nutrients while keeping all other materials out.

The way that osmosis works can easily be demonstrated with a carrot. If the carrot is placed in a saline solution, water flows out of the carrot cells to dilute the saline solution, and the carrot shrivels. If the carrot is now placed in pure water, the water flows into the carrot and it swells.

In nature, the concentration of saline solution is critical. In animals, if it exceeds 0.9% then there is a real chance that the effect of osmosis will be so powerful that the cells will be ruptured. Plants, by contrast, have much stronger cell walls and can put up with a high saline solution.

▲▶ If you put a carrot in a strong brine solution, natural osmosis will work on the cells of the carrot tissue and the carrot will shrivel up. This gives a very dramatic impression of the power of osmosis. In the picture above, the carrot has been placed in distilled water, while the picture on the right shows a carrot that has been placed in strong brine.

Plants that thrive in salt water

Many plants thrive in salt water. Indeed, plants are common in the sea. Algae, the microscopic green plants that float on the surface of the sea, are at the bottom of the sea food chain. Other large plants, such as seaweeds, are common sea water plants.

▲ Mangroves thrive in coastal conditions, as they are able to withstand salt.

◀ Marine fish are able to resist the osmotic effect of salty sea water.

Sea water life

For animals and plants to be able to survive they need to have adapted. All ocean creatures, for example, have adapted so that they can tolerate much higher concentrations of sodium than fresh water plants and animals.

Salt lake life

The high salt concentration of salt lakes makes it especially difficult for most aquatic life. However, some species of shrimp and algae can thrive, and these form the food of salt lake birds such as the flamingo of East Africa.

Desert life

Desert water and soils have a very high level of salt. Only a few species have adapted to life under such harsh conditions. One of the most common is called the saltbush. A form of eucalyptus tree has also adapted, producing a special kind of vegetation, known in Australia as "mallee". The saltbush scrubland of America is a further example.

Coastal life

Marram grass is a salt-tolerant land plant, found worldwide, that grows on sand dunes by the coast. In tropical regions, mangrove forests thrive in salt water, often forming impenetrable coastal forests that protect the coast from the erosive energy of the waves.

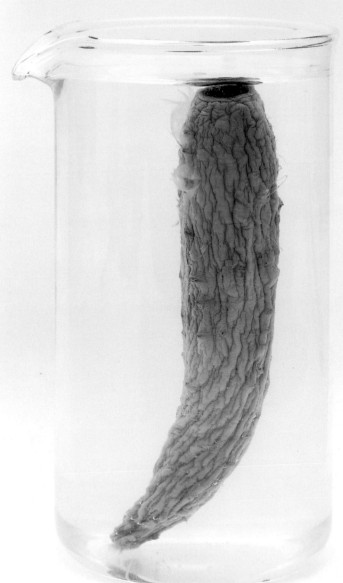

Sodium and the human body

Sodium, in the form of salt, is present in many of the foods we eat. In general, if we eat animal products such as milk, fats or meat, we do not need to add extra salt to any of our foods.

Why is it, then, that our tongues have special areas that detect salt? And why do we like the taste of salt so much? The answer lies in part in the role of osmosis.

Salt in the blood

Biologically, blood is a saline solution, that is, it is mostly water with salt dissolved in it. The red blood cells, present in the saline solution, which give blood its familiar appearance, are a relatively small part of the liquid we call blood.

The main use of salt in the body is to control the way water is taken up and expelled from the organs. Salt is also important in making sure that nerve cells work properly.

Blood contains more salt and less water than digested food, water, and other energy-giving molecules. These are then are attracted towards the blood as it courses through vessels that line the digestive tract. The blood then carries nutrients to all the cells in the body.

Saline drip for emergencies

When people lose blood, the first thing that a doctor does is to give them an intravenous injection of water and salt, known as a saline drip. A saline drip contains no blood cells, so it does not have to be matched to the blood group and can be used on anybody.

Using a saline drip will ensure that there is enough liquid in the bloodstream until alternative supplies of matching blood can be found.

◄ Oral rehydration salts
In some circumstances, such as when the body is affected by a disease that causes diarrhoea, large amounts of fluids are lost. The body does not have time to filter out the salts it needs and these are carried away with the fluids. To help people to recover from these illnesses, the body needs not just fluids, but also a balanced range of salts, so that it can quickly regain the ability to absorb nutrients and energy from food.

To help the body regain the salts, patients are given oral rehydration therapy, consisting of a balanced mixture of salts dissolved in water. This cheap and effective cure saves millions of lives each year.

Kidneys

The kidneys have the task of filtering out waste products from the body. The kidneys keep most of the sodium that is in the body, but up to five grams of sodium (a small teaspoonful) are excreted each day. Because sodium is so vital to the functions of the body, animals must make up this lost sodium in the food they eat. Fortunately, a normal diet provides this amount.

Sometimes kidneys fail, causing toxic wastes and fluids to build up in the blood. When this happens, patients are put on a blood-filtering, or dialysis, machine.

The first artificial kidney used a drum with a semipermeable membrane made from cellophane. The drum was bathed in a saline solution, while blood entered the drum.

The saline solution attracted out the wastes from the drum by osmosis, leaving the blood clean. Since then, major improvements to the process have been developed, but the principle of cleaning the blood remains the same.

▼ As this patient waits for treatment in a hospital, a saline drip is being used to regulate the blood.

digestive tract: the system of the body that forms the pathway for food and its waste products. It begins at the mouth, and includes the stomach and the intestines.

molecule: a group of two or more atoms held together by chemical bonds.

osmosis: a process where molecules of a liquid solvent move through a membrane (filter) from a region of low concentration to a region of high concentration of solute.

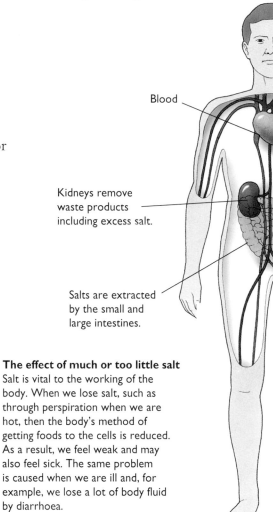

Blood

Kidneys remove waste products including excess salt.

Salts are extracted by the small and large intestines.

The effect of much or too little salt
Salt is vital to the working of the body. When we lose salt, such as through perspiration when we are hot, then the body's method of getting foods to the cells is reduced. As a result, we feel weak and may also feel sick. The same problem is caused when we are ill and, for example, we lose a lot of body fluid by diarrhoea.

As we lose salt, our body reacts by wanting to get more salt. This is why we sometimes crave salty things.

We can also become ill if we take in too much salt. Salt controls the pressure in the bloodstream, and too much salt can lead to high blood pressure and the risk of damaging the heart. Normally, we get rid of excess salt by sweating it off through the pores in the skin. This is the reason your skin sometimes tastes salty if you lick it. But if we take in far too much salt (say if we were to eat a tablespoon of salt), the body reacts to this by ejecting the salt through vomiting.

Salt pollution

Salt can build up in soils making it impossible for crops to grow. Soil scientists called the build up of unwanted salt salinisation. It is a form of salt pollution.

All waters contain dissolved salts. In most cases these salts are good for plants, providing essential minerals for growth. However, in dry areas salts can build up if the land is not irrigated properly.

▲ Under normal conditions, rainwater flushes out the salt from fields. In a desert, there is not enough rainfall to do this. Thus, the only plants that will grow are those that will tolerate salty soils.

Why soils become salty

Salt occurs in nature from rainwater, the decay of dead plants, river water that soaks into the soils after floods, and the weathering of rock. The salt is absorbed by humus and clay particles in the soil to provide a reserve bank of nutrients for plants to use in their growth. However, too much salt can cause problems. Thus in most parts of the world, bouts of heavy rainfall and floods that soak right through the soil are vital in helping to flush surplus minerals away.

In dry regions, farmers disturb the natural balance by irrigating the land for long periods each year. Many use water drawn from underground supplies that are nearly saturated with salts. The farmers may be tempted to save precious water by putting only enough water on the soils to provide sufficient moisture for the plant roots. As a result the surplus minerals never get flushed away. Indeed, they are often drawn back up to the surface as the water in the soil evaporates.

You can see salt-polluted, or saline, soils because they have white salt crystals on their surface.

▶ The irrigation water applied to fields in arid areas often has been pumped from underground supplies, or aquifers, and these can have a very high concentration of salt.

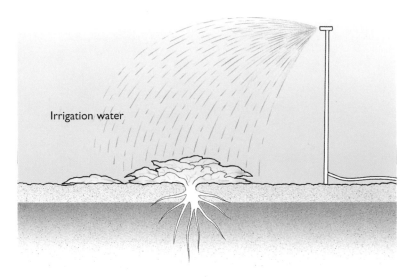

Irrigation water

irrigation: the application of water to fields to help plants grow during times when natural rainfall is sparse.

salinisation: the concentration of salts, especially sodium chloride, in the upper layers of a soil due to poor methods of irrigation.

▲ Under irrigation, it is vital that the salt is flushed out of the soil. This is because water used for irrigation will often have come from groundwater supplies and in its long passage through rocks it will have picked up a higher salt content than rainwater.

Farmers are often tempted to put on just enough water to feed the roots with the water they need, but this is a very short-sighted policy because unless the salt is flushed away it will pollute the soil.

A good irrigation system will have deep drains to make sure the water is carried from the soil as well as sprinklers to apply the water.

▲ In some places irrigation waters are so salty that significant amounts of salt are left in the bottom of a beaker after the water has been evaporated.

Soils most at risk

It is likely that many areas of desert were once fertile before salts built up in them because of poor irrigation. Modern areas badly affected include the Murray–Darling basin of Australia, the Indus Valley of Pakistan and the Colorado Basin of North America.

◄ Soils that have suffered salt damage may not be suitable for use for over 20 years even if irrigated properly thereafter.

Salt solution as a conductor

A compound is a combination of two or more elements held together by small electrical charges. These charges work much like the electrical charge that is formed when you rub a balloon. The electrical charge created makes it "stick" to your clothes.

Charged particles are called ions, so a charged particle of sodium is called a sodium *ion*. Sodium ions have a positive charge and chloride ions have a negative charge. As opposites attract, the two types of ion are attracted to each other and bond together as sodium chloride.

Many compounds, including sodium chloride, dissociate (break apart) in water, however. When this happens, the charged nature of the elements becomes obvious. Water containing sodium and chloride ions is known as an electrolyte. An electrolyte can allow an electrical current to flow through it.

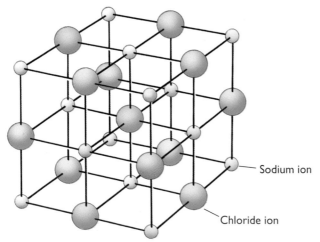

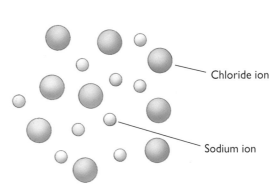

▲ A crystal of salt is a solid made up of sodium and chloride ions firmly bonded together.

▲ When salt is put in water, the bonds break down (the salt dissociates) and the sodium and chloride ions move freely in the water. This is why the salt solution is a good electrolyte.

Salt as an electrolyte

To demonstrate how salt behaves as an electrolyte, salt is placed in water and a battery is connected to a circuit like that shown here.

Because the salt water conducts electricity, the bulb will light. This happens because the sodium and chlorine ions are free to move. When a voltage is applied, the negative ions flow towards the positive side of the battery, and the positive ions move towards the negative side of the battery. If the ions were not free to move, they could not complete the circuit in this way. Pure water, for example, is a poor electrical conductor and solid sodium chloride will not conduct electricity at all.

bond: chemical bonding is either a transfer or sharing of electrons by two or more atoms.

electrolysis: an electrical–chemical process that uses an electric current to cause the break up of a compound and the movement of metal ions in a solution. The process happens in many natural situations (as for example in rusting) and is also commonly used in industry for purifying (refining) metals or for plating metal objects with a fine, even metal coating.

electrolyte: a solution that conducts electricity.

ion: an atom, or group of atoms, that has gained or lost one or more electrons and so developed an electrical charge.

solution: a mixture of a liquid and at least one other substance (e.g. salt water). Mixtures can be separated out by physical means, for example by evaporation and cooling.

◄▼ When copper electrodes connected to a battery are immersed in a salt solution, a current will flow. Thus metal will leave the anode of the cell and migrate through the water to be redeposited on the cathode. You can see this effect if the cell is left running for some time. The colour of the solution begins to change, showing that copper has been dissolved from the anode and transferred to the water. The copper anode also becomes brighter as the metal is lost from the surface.

Also...

Molten salt conducts electricity. If salt is heated until it melts and an electric current flows through it, then the sodium ions will move to the cathode (where solid sodium metal will accumulate) while chloride ions will move to the anode where they are given off as gas. This process can be used in industry to produce both chlorine gas and sodium metal.

Properties of salt

Salt, sodium chloride, is use widely for cooking, preserving and also for preventing roads from icing up. This is because it has a number of unusual properties, as described on this page.

How salt changes the way food cooks

A $10°C$ rise in temperature doubles the rate of chemical reaction. Salt raises the boiling point of water by several degrees and so the chemical reactions that occur when food is cooked happen at a higher temperature, and therefore faster. The higher cooking temperature also tenderises the food more effectively than when cooked at a lower temperature.

Salt also affects the taste of the food because we have taste buds that can detect salt. However, this is a separate effect.

▲▶ Salty water boils at a much higher temperature than pure water as shown by this simple demonstration with salt, water, a beaker and a thermometer.

Ice preventer

The temperature at which water turns from a liquid to a solid is $0°C$. Any impurity in the water will, however, lower the freezing point. Salt is used because it is a convenient and cheap material that will dissolve in water and add the necessary impurities. The more salt that dissolves in water, the lower its temperature can fall without freezing. Rivers, for example, contain less salt than the sea and so freeze at a higher temperature.

Icebergs can co-exist with sea water in cold environments. The iceberg is made of pure rainwater which freezes at a higher temperature than the salty sea water. In fact, because the seas near Antarctica have very little rainfall, the sea water is particularly salty and so freezes well below $0°C$.

Preserving

In general, the food that we eat is harvested at one time of the year and eaten over the subsequent weeks and months. In the past, when people did not have freezers, they had to use chemical means to preserve food. One way was to use salt.

Salt preserves food by binding itself to the water in food so that micro-organisms cannot use it. It also sucks water out of the bacteria, killing them.

There are a number of ways that food can be preserved by salt. Fish and some meats are packed in salt, and have salt rubbed into their skins.

Curing is a process of adding salt to meat. It is the same as pickling. Bacon is the most common cured meat, produced by salting pork bellies, ham and shoulder.

The salt is first rubbed into the meat and then brine is injected into the meat using a tool with many needles. This gets the brine inside the meat in an even manner. It is then sometimes hung in wood smoke for a number of days.

Pickling is a way of preserving food using salt and water – brine. It suits some vegetables and meats better than others. Common foods that have been pickled in brine include sauerkraut (pickled cabbage), and gherkins (pickled cucumbers). Other foods that can be pickled include sardines, mackerel and eggs.

Not only does the salt prevent decay, it also sucks the moisture from the food, drying it out, although this is different from the process called dehydration. This is why many salted foods have to be soaked in water for long periods in order to get the excess salt out before they can be eaten and to make the food moist enough for it to be chewed.

Also…

The principle at work in taking the water out of material by packing or pickling them in salt is osmosis. This causes water to move through a semipermeable membrane (filter) such as skin if the concentration of salt on one side of the membrane is greater than on the other side. Pickling materials in a brine solution means that the brine has a more concentrated solution than the food, hence water moves from the food into the brine.

dissolve: to break down a substance in a solution without a resultant reaction.

freezing point: the temperature at which a substance changes from a liquid to a solid. It is the same temperature as the melting point.

osmosis: a process where molecules of a liquid solvent move through a membrane (filter) from a region of low concentration to a region of high concentration of solute.

Chemicals from salt

Salt is one of the most important chemicals for modern industry. It is the starting material from which many other chemicals are made.

One of the most important, efficient and elegant processes for getting products from salt was developed by Hamilton Castner and Karl Kellner at the end of the 19th century. The process involves passing an electric current through brine (salt solution).

During the reaction, which takes places in a container called a diaphragm cell, a number of vital chemicals are produced, as shown by the reaction on this page.

A different process, the Downs process, is used to obtain sodium.

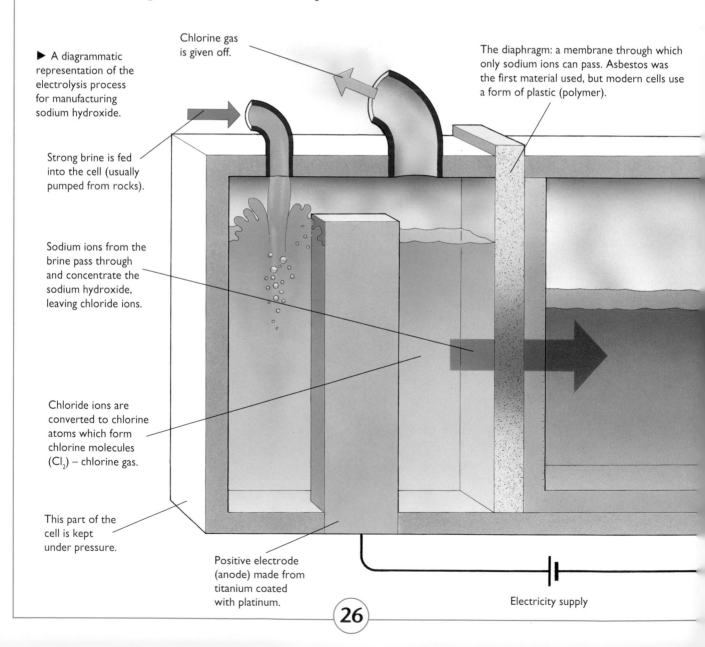

► A diagrammatic representation of the electrolysis process for manufacturing sodium hydroxide.

Chlorine gas is given off.

The diaphragm: a membrane through which only sodium ions can pass. Asbestos was the first material used, but modern cells use a form of plastic (polymer).

Strong brine is fed into the cell (usually pumped from rocks).

Sodium ions from the brine pass through and concentrate the sodium hydroxide, leaving chloride ions.

Chloride ions are converted to chlorine atoms which form chlorine molecules (Cl_2) – chlorine gas.

This part of the cell is kept under pressure.

Positive electrode (anode) made from titanium coated with platinum.

Electricity supply

- **anode**: the negative terminal of a battery or the positive electrode of an electrolysis cell.

- **brine**: a solution of salt (sodium chloride) in water.

- **cathode**: the positive terminal of a battery or the negative electrode of an electrolysis cell.

- **ion**: an atom, or group of atoms, that has gained or lost one or more electrons and so developed an electrical charge.

- **membrane**: a thin flexible sheet. A semipermeable membrane has microscopic holes of a size that will selectively allow some ions and molecules to pass through but hold others back. It thus acts as a kind of sieve.

- **molecule**: a group of two or more atoms held together by chemical bonds.

EQUATION: Electrolysis of a salt solution

Sodium chloride + water ⇨ sodium hydroxide + chlorine + hydrogen

$$2NaCl(aq) + 2H_2O(l) \xrightarrow{\text{electrical energy}} 2NaOH(aq) + Cl_2(g) + H_2(g)$$

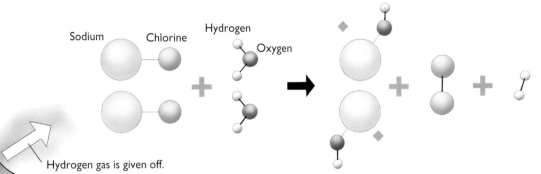

Sodium   Chlorine   Hydrogen   Oxygen

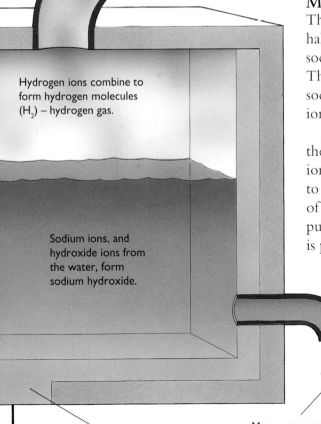

Hydrogen gas is given off.

Hydrogen ions combine to form hydrogen molecules (H_2) – hydrogen gas.

Sodium ions, and hydroxide ions from the water, form sodium hydroxide.

The negative electrode (cathode) is a perforated steel box.

More concentrated sodium hydroxide is produced.

Manufacture of sodium hydroxide

This method uses an electrolytic cell that is divided in half by a semipermeable membrane. Brine, containing sodium and chloride ions, is pumped into the cell. The semipermeable membrane is designed so that sodium ions can pass through, but the larger chloride ions cannot.

An electrical current attracts the sodium ions through the semipermeable membrane. The surplus chloride ions in the left-hand part of the cell combine together to form chlorine gas molecules, which then bubble out of the brine and are collected. The used up brine is pumped out of the cell and more concentrated brine is pumped in.

A supply of sodium hydroxide is pumped through the right-hand side of the cell. The sodium ions combine with the water to form more sodium hydroxide, thus increasing its concentration. At the same time, surplus hydrogen ions in the water combine together to form hydrogen gas, which bubbles out of the solution and is collected.

The diaphragm also prevents the hydrogen and chlorine gases from mixing and allows them to be collected separately.

Manufacturing sodium carbonate

Sodium carbonate (a chemical needed for soap and glass–making) was originally made by soaking plants in water. However, this used a huge amount of plant material and produced very little chemical.

The first scientifically designed chemical process to make sodium carbonate was invented by Nicolas Leblanc in 1789. He poured sulphuric acid on to common salt and then treated the result with limestone and charcoal.

In 1861, a less complicated and far more efficient process was invented by the Belgian scientist, Ernest Solvay. The Solvay process begins with a strong brine solution, through which ammonia and carbon dioxide are then bubbled. The three substances then combine to produce sodium bicarbonate (baking soda). When this is heated, it forms sodium carbonate.

How the Solvay process works

The process uses brine pumped from underground salt beds, which is reacted with ammonia and carbon dioxide. Limestone quarried from rocks and ammonia are needed to begin the process, but as you will see from the diagram opposite, once the process is under way it produces all the chemicals it needs except for brine.

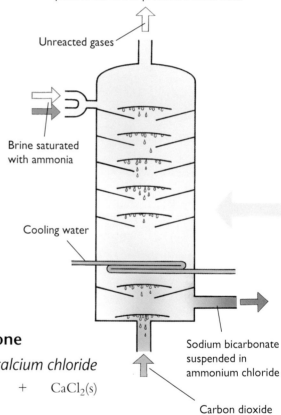

▼ The Solvay tower. Water is pumped in to cool the tower as the reactions taking place in the tower produce a lot of heat.

Unreacted gases

Brine saturated with ammonia

Cooling water

Sodium bicarbonate suspended in ammonium chloride

Carbon dioxide

EQUATION: Overall reaction of salt and limestone

Sodium chloride + limestone ⇨ sodium carbonate + calcium chloride

$2NaCl(aq)$ $+$ $CaCO_3(s)$ $⇨$ $Na_2CO_3(s)$ $+$ $CaCl_2(s)$

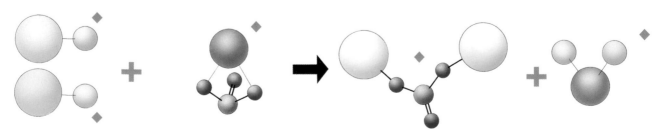

EQUATION: Reaction of salt with ammonia and carbon dioxide

Sodium chloride + ammonia + carbon dioxide + water ⇨ sodium bicarbonate + ammonium chloride

$$NaCl(s) \quad + \quad NH_3(g) \quad + \quad CO_2(g) \quad + \quad H_2O(l) \quad ⇨ \quad NaHCO_3(s) \quad + \quad NH_4Cl(s)$$

EQUATION: Reaction on heating sodium bicarbonate

Sodium bicarbonate ⇨ sodium carbonate + water + carbon dioxide

$$2NaHCO_3(s) \quad ⇨ \quad Na_2CO_3(s) \quad + \quad H_2O(l) \quad + \quad CO_2(g)$$

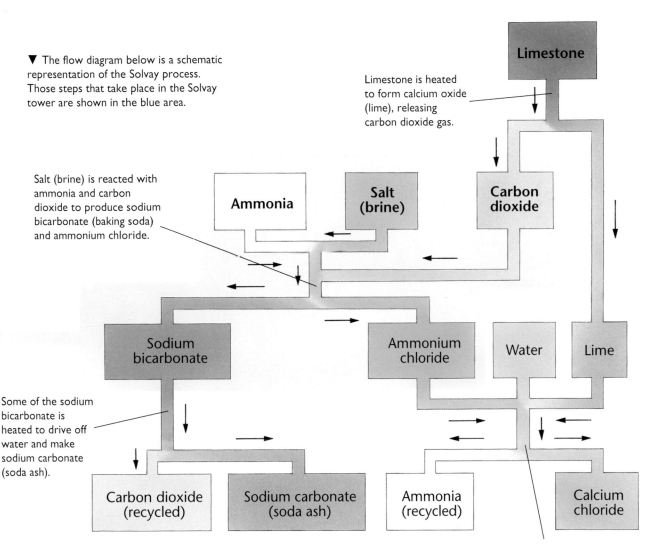

▼ The flow diagram below is a schematic representation of the Solvay process. Those steps that take place in the Solvay tower are shown in the blue area.

Limestone is heated to form calcium oxide (lime), releasing carbon dioxide gas.

Salt (brine) is reacted with ammonia and carbon dioxide to produce sodium bicarbonate (baking soda) and ammonium chloride.

Some of the sodium bicarbonate is heated to drive off water and make sodium carbonate (soda ash).

Ammonium chloride is reacted with calcium oxide (lime) to form calcium chloride, water and ammonia. This part of the process allows the recovery of the ammonia for further use.

The products
The products are sodium bicarbonate (baking soda), sodium carbonate (soda ash, needed for soap and glass-making), carbon dioxide gas (which is recycled and can be used for other industries) and calcium chloride (the only "waste" product, but which is mixed with salt and then used as a chemical for melting snow and settling dust on roads).

Sodium carbonate and bicarbonate

Soda is a name used to refer to a number of kinds of sodium compound. Sodium carbonate is also called soda ash because it was once obtained from the ashes of plants. Sodium bicarbonate is called baking soda, so named for its main use.

▼ Crystals of sodium carbonate are used as bath salts. Sodium carbonate acts as a water softener. The bath salt crystals naturally contain a large proportion of water and so they will dissolve in bath water readily.

Sodium carbonate

Sodium carbonate is mainly used in making glass. Specifically, it is used to make the molten glass less viscous and thus easier to handle.

Sodium carbonate is also used to treat sewage and to soften water. It can also be used in the dyeing and leather-tanning industries. When combined with arsenic, sodium carbonate was an important pesticide; it also protects against bacteria.

Sodium bicarbonate in baking

Baking powder is mostly sodium bicarbonate (baking soda) together with cream of tartar (a natural organic acid), and small amounts of starch. It is used to make cakes rise (or leaven) as they are cooked.

Also...

Although yeast will also cause cakes to rise, its biological action is less easy to control than the straightforward chemical reactions in baking powder.

◄ **How baking powder works**
Cakes rise because of the action of bubbles of carbon dioxide gas. Baking soda is used to produce the carbon dioxide gas.

When baking powder is added to a wet cake mix, it changes from a solid powder to become liquid. The soda reacts with the tartaric acid crystals to release carbon dioxide as bubbles. These bubbles are also released as the baking soda decomposes on being heated.

The faster the gas is released, and the bigger the bubbles of gas produced, the more the cake will rise, and the fluffier the result. The baking powder mixture is designed to control the rate of gases before and during cooking.

EQUATION: Producing carbon dioxide by heating baking soda

Sodium bicarbonate ⇨ sodium hydroxide + carbon dioxide + water vapour

$$2NaHCO_3(s) \quad \Rightarrow \quad Na_2CO_3(s) \quad + \quad CO_2(g) \quad + \quad H_2O(g)$$

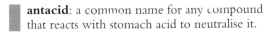

> **antacid**: a common name for any compound that reacts with stomach acid to neutralise it.

◄ This demonstration of the reaction of baking powder with vinegar gives an insight into the reaction of an antacid such as baking powder in the stomach. The equation for this reaction is shown below

EQUATION: Baking soda and hydrochloric acid

Stomach acid + sodium bicarbonate ⇨ sodium chloride + water + carbon dioxide

$$HCl(aq) \quad + \quad NaHCO_3(s) \quad ⇨ \quad NaCl(aq) \quad + \quad H_2O(l) \quad + \quad CO_2(g)$$

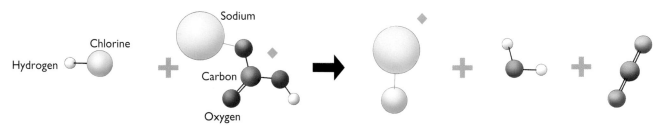

◄ An antacid tablet is placed into dilute hydrochloric acid as a demonstration of the reaction that takes place inside the stomach. The bubbles of carbon dioxide gas are generated with sufficient vigour to break up the tablet, increasing the area exposed to the acid and thereby speeding up the reaction.

Sodium carbonate and bicarbonate as antacids

Sodium carbonate and bicarbonate are used as antacids to balance the excess of hydrochloric acid that sometimes builds up in the stomach. The sodium carbonate and bicarbonate react with acids to produce a harmless chalk-like material that can be taken away during digestion. Their only side effect is that they give off carbon dioxide gas – which is why people burp when taking this form of antacid.

Sodium bicarbonate as a neutralizer

Baking powder is a mildly alkaline substance that can be used to counteract the sting of a bee.

When a bee stings, it injects an acid just below the skin. This causes the body to react violently, producing pain and severe itching.

The skin is not equipped to send large amounts of an alkali to counteract the acid sting. But baking soda is just one of a number of household substances that can be used to give relief.

If the baking powder is rubbed into the site of the sting, it will react with the acid, neutralising it and making the sting less painful.

Sodium hydroxide

Sodium hydroxide, commonly known as caustic soda, is used in a wide range of chemical processes, from soap-making (see page 34), to manufacturing dyes and cosmetics, paper and petrol. It can also be used as an environmentally friendly agent because it is able to remove sulphur-containing gases that might otherwise form acid rain.

Sodium hydroxide solution is colourless and reacts with many metal compounds or acids. The action of sodium hydroxide on aluminium is quite remarkable, as shown below, which suggests that you don't clean aluminium pans with caustic soda.

Sodium hydroxide is made from brine by passing an electric current through it as shown on page 26. Alternatively it can be made by reacting calcium hydroxide with sodium carbonate (the product of the Solvay process, page 28).

Properties of sodium hydroxide

Sodium hydroxide is a base, and as such it will produce an alkaline reaction when tested with a chemical indicator.

Sodium hydroxide is one of the few bases that are soluble in water. This makes it a very useful chemical. However, sodium hydroxide is a caustic material and will burn the skin and corrode some metals.

▼ The demonstration below shows how sodium hydroxide corrodes aluminium.

❶ The small aluminium container used in the demonstration.

❷ Sodium hydroxide is poured in and a reaction immediately begins.

❸ Within a few minutes the base of the container has reacted and dissolved.

EQUATION: Dissolving aluminium in sodium hydroxide

Aluminium (solid metal) + sodium hydroxide (liquid) ⇨ sodium aluminate (liquid) + hydrogen

$$2Al(s) + 2NaOH(aq) + 6H_2O(l) \Rightarrow 2NaAl(OH)_4(aq) + 3H_2(g)$$

◀ Sulphur-containing exhaust gases are removed from the tall chimneys of this power station.

Sodium hydroxide as a pollution-preventing agent

Sodium hydroxide is used to neutralise weak acids. For example, it is used in the petroleum industry to remove sulphur-containing impurities such as hydrogen sulphide gas. It is also used to remove sulphur dioxide from the exhaust gases of some power stations. This is known as "scrubbing". The sulphur is used to make sulphuric acid.

EQUATION: Removing sulphur-containing impurities such as hydrogen sulphide from petroleum with sodium hydroxide

Hydrogen sulphide + sodium hydroxide ⇨ sodium sulphide + water

$$H_2S(g) \quad + \quad 2NaOH(aq) \quad ⇨ \quad Na_2S(l) \quad + \quad 2H_2O(l)$$

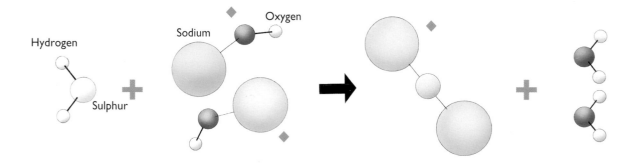

EQUATION: Removing sulphur dioxide from power station exhausts with sodium hydroxide

Sulphur dioxide + sodium hydroxide ⇨ sodium sulphate + water

$$SO_2(g) \quad + \quad 2NaOH(aq) \quad ⇨ \quad Na_2SO_3(s) \quad + \quad H_2O(l)$$

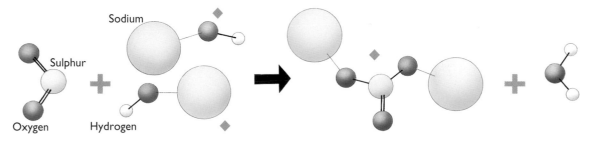

Sodium hydroxide as a household chemical

Sapo is the Latin word for soap. Saponification therefore, means making soaps.

Saponification is a natural reaction of sodium hydroxide and any animal fat or vegetable oil. This is a most valuable reaction because it means that sodium hydroxide can be used to make soap and it can also be used in drains and other places to destroy unwanted build-up of fat or oil.

Caustic soda and water

If you put crystals of sodium hydroxide into water an immediate reaction occurs, releasing considerable amounts of heat. The heat released can bring the water close to boiling within seconds.

Making a solution of sodium hydroxide therefore creates a very hot, very strong alkaline substance, and one that must be treated with care and respect, especially as it is commonly kept in many household cupboards.

▼ When sodium hydroxide crystals are added to water they create a hot, caustic solution. The picture on the left shows a beaker of water before the sodium hydroxide crystals were added. The thermometer on the right shows the temperature increase as the crystals were added. This is an exothermic reaction, i.e. it produces heat.

◄ Sodium hydroxide crystals. When added to water they create a hot, caustic solution.

▼ Oven-cleaning pads are made with caustic soda because a very strong reaction is needed. The reactive face of oven pads should not be handled because caustic soda will react with skin to form soap!

EQUATION: Caustic soda and water

Sodium hydroxide + water ⇨ dilute sodium hydroxide

$NaOH(s)$ + $H_2O(l)$ ⇨ $NaOH(aq)$

heat given off

Caustic soda and blocked drains

When caustic soda is added to water and then put down a drain that has been blocked by fats and oils, the reaction of water and the sodium hydroxide produces a hot alkaline solution. Chemical reactions occur much faster when substances are hot rather than cold.

When hot sodium hydroxide reacts with animal fats, it breaks up the long chain-like molecules of the fat or oil, making short pieces that can easily be washed away.

EQUATION: Caustic soda and fat

Sodium hydroxide   +   fat   ⇨   soap   +   glycerol

$3NaOH(aq) + (C_{17}H_{35}CO_2)_3C_3H_7(s)$ ⇨ $3(C_{17}H_{35}CO_2)Na(aq) + C_3H_7(OH)_3(aq)$

Soap

The reaction of sodium hydroxide with fat can be used to make soap. Soap is a waxy solid used in cleaning. It is different from a detergent, whose chemical action is based on petroleum products.

Soap-making is one of the most ancient chemical reactions, used for at least two thousand years. It was used not just for cleaning, but also for keeping skin healthy. It was noticed that when people washed in soap, they were much less prone to skin diseases.

Saponification

Soap can be made easily in the laboratory. In the flask shown below oil has been gently added to a volume of sodium hydroxide solution in water. It forms a yellow layer on the top of the hydroxide.

The flask and its contents are boiled with a condenser over the top of the flask to make sure that nothing evaporates during the boiling process. After a while the whole of the liquid takes on an opaque appearance and bubbles begin to appear on the surface. This is liquid soap.

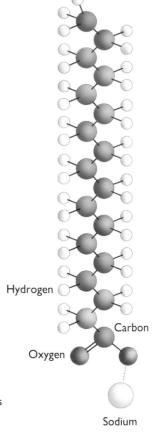

▼ This end of the sodium stearate molecule is repelled by water and molecules of the same type but is attracted to grease. The chemical formula for this molecule is $C_{17}H_{35}CO_2Na$.

Hydrogen

Carbon

Oxygen

Sodium

Oil

Sodium hydroxide solution.

The round bottomed flask is heated and the contents boil.

After boiling, the solution becomes opaque. Soap has been produced.

Also...

When soap is used in hard-water areas the water often develops a scum that forms a deposit on bath tubs and on clothes. This scum is formed because the soap molecule is attracted to the calcium and magnesium salts in hard water and it combines with these minerals to form a white precipitate rather than dissolving grease away.

Detergents are used for washing to prevent this problem; however, detergents are too harsh for use in washing skin. The scum can be counteracted by using a water softening agent (such as bath salts, sodium carbonate) in the bath or in the soap.

How soap works

When soap is mixed with water, it breaks up into lollipop-shaped molecules. One end is attracted to water, while the other end repels water. This means that the water-hating (hydrophobic) ends are attracted to the surfaces of all objects, because this keeps them away from the water.

When a soap molecule gets close to a piece of dirt, the water-hating end sticks to it. If the dirt is rolled around, such as it might be the case when we wash our hands, there is plenty of opportunity for soap molecules to stick all round the dirt.

In this way the molecules surround the dirt, stopping it from sticking back to the surface we are trying to clean. At the same time, the water loving ends of the molecules are attracted to the water. As a result the whole molecule, and the dirt are pulled away from the surface. In this form the dirt cannot settle out and stick to a surface and so it is easily washed away.

Soap molecules cannot clean all by themselves. They need to be pushed around, so they wrap up the dirt molecules as they are moved about. This is why we have to rub our hands or why a washing machine has to agitate to get the dirt off.

detergent: a petroleum-based chemical that removes dirt.

saponification: the term for a reaction between a fat and a base that produces a soap.

❸▼ The heads of the detergent molecules are repelled by one another and so combine, with some agitation, with the water. By holding your hand under the tap, the grease is broken up into smaller particles and can be washed away from the surface.

❷▼ The tails of the detergent molecules are attracted to the grease on the surface. The water-loving (hydrophilic) heads are left in the water.

❶▼ The detergent is added to the water.

Droplet of soap

Soap molecule

Water

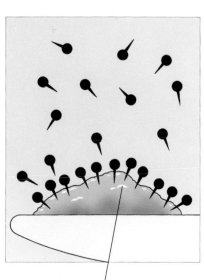

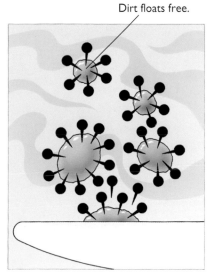

Dirt floats free.

Grease

Surface with greasy dirt, skin, clothes, etc.

Soap molecules embed themselves in the grease.

Sodium light

When sodium is heated it burns with a brilliant orange flame. This property can be seen in almost every city in the world, where it is used to provide the bright orange illumination for city streets.

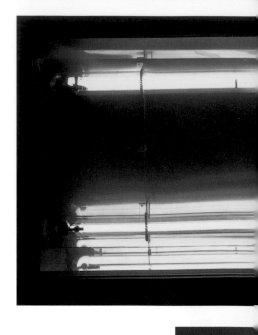

Discharge tubes

The way sodium is used for lighting is not unlike the way light is produced in a fluorescent tube in your home. The tube contains a gas at a high pressure. In the case of sodium it is called sodium vapour.

At both ends of the tube are electrodes, pieces of metal that can introduce electricity inside the tube.

When a very high voltage is applied to the tube, the vapour begins to glow. High pressure sodium vapour lamps give out a brilliant yellow–orange glow.

Living with yellow–orange light

The light produced by a sodium lamp is of one colour. This is quite unlike the light produced by the Sun, which contains many colours mixed together to produce white light.

Yellow–orange light makes many objects look dull and flat. To make the light more acceptable, some sodium vapour tubes have a coating of a fluorescent material (just like fluorescent tubes at home). This changes the orange light to a slightly whiter light.

▶ A piece of sodium being heated in a Bunsen burner flame shows the characteristic yellow–orange light of sodium.

◀▼ A tunnel is lit by sodium discharge tubes. The picture on the left shows the detail.

fluorescent: a substance that gives out visible light when struck by invisible waves such as ultraviolet rays.

vapour: the gaseous form of a substance. For example, water vapour is the gaseous form of liquid water.

▼ Sodium lights are only one of a range of elements used to create lighting. Many "neon" lights are elemental colours; for example, neon gives out a red light in a discharge tube. Most metal filaments in incandescent bulbs produce a yellow–white light. A panorama of a city at night will show many of these colours, but almost certainly sodium lights will dominate as they are efficient and relatively cheap.

Potassium

Potassium is found in the sea, in rocks and in plants. It was first obtained from the ashes of plant, as potassium carbonate, known as potash. However, the largest amount of potassium is in sea water. There is nearly half a kilogram of potassium oxide in each cubic metre of sea water and much of this can be recovered in salt pans (along with sodium chloride, common salt) as the water evaporates away. Potassium salts are also found among deposits in desert lakes and in ancient desert rocks.

Potassium is also found in many volcanic rocks. The pink or white feldspar crystals in granite, for example, are compounds of potassium. In turn, when feldspar weathers to clay, the potassium becomes part of the clay particles that make up the world's soils.

Potassium compounds

Potassium reacts readily with the halogens to form potassium halides and with oxygen to form potassium oxide and potassium peroxide.

Potassium is vital to plant growth and large amounts are used in the form of fertilisers such as potassium nitrate and potassium sulphate.

Potassium phosphate used to be widely used in detergents. However, because the phosphate then found its way through the sewage systems into rivers and caused water pollution, it is now rarely used for this purpose.

Potassium carbonate (potash) is used in glass-making.

▼ Potassium salts were traditionally obtained from the ashes of trees and other plants. The ash yields potassium carbonate contaminated with a wide variety of other compounds. Potassium is now recovered from coastal salt beds at the same time as brine, and also from underground rock deposits.

A mixture of 75% potassium nitrate, 15% carbon (charcoal), and 10% sulphur is the "black powder" used as gunpowder for more than 2200 years.

▲▶ Gunpowder, which contains potassium, is often used in modern fireworks to give a dramatic explosion.

This is a feldspar crystal in a piece of granite. Crystals of volcanic rock like this are the origin of the world's sodium and potassium compounds.

Potassium and the human body

Potassium, like sodium, is essential to all life. Nearly all potassium is used inside the cells, making sure that the acid levels and fluid pressures remain normal. Potassium is particularly used in nerves and muscles.

The kidneys control the amount of potassium in the body. If the kidneys do not maintain certain levels, then heart problems can occur. In general, potassium is found in all foods, so that apart from when kidney diseases occur, the human body has no problem acquiring all the potassium it needs.

▼ The amount of potassium salts in water can be seen in some places where the waters are potassium rich. These are called soda lakes. This is a picture of potash salts encrusting the side of one such lake.

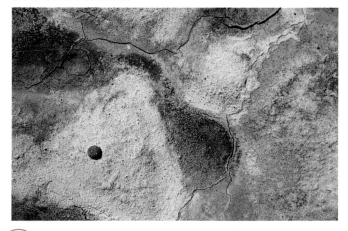

Potassium reactions

Potassium is highly reactive. It not only reacts violently with water, it also forms a wide variety of compounds. In many ways these compounds are similar to those of sodium. However, the reactions of many potassium salts will be different from those of similar salts of other metals because the potassium plays no further part in reactions and therefore its salts are relatively unreactive (stable). Thus potassium nitrate will only decompose if heated well above its melting point. It then reluctantly releases oxygen gas and leaves an even more stable potassium salt.

▲▼ When potassium metal is added to water it reacts even more vigorously than sodium (compare the reaction to that shown on page 6). As the potassium burns, it produces a violet flame. The solution produced is caustic potassium hydroxide, which turns the indicator in the solution dark pink.

EQUATION: Potassium and water

Potassium + water ⇨ potassium hydroxide + hydrogen

$$2K(s) \quad + \quad 2H_2O(l) \quad ⇨ \quad 2KOH(aq) \quad + \quad H_2(g)$$

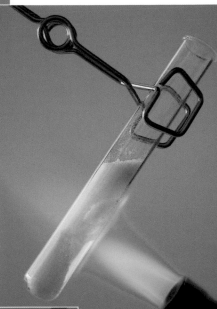

EQUATION: Heating potassium nitrate

Potassium nitrate ⇨ potassium nitrite + oxygen

$$2KNO_3(s) \quad \Rightarrow \quad 2KNO_2(s) \quad + \quad O_2(g)$$

◄◄ When potassium nitrate is melted, it produces a greenish yellow liquid (potassium nitrite) and gives off oxygen gas. Potassium nitrite is not decomposed by heating; and when the heat is removed, it eventually cools to a white solid.

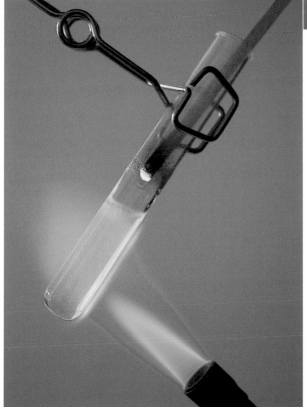

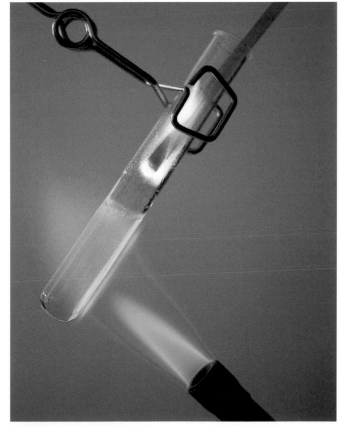

▲► The presence of oxygen is revealed by the rekindling of a smouldering splint.

Key facts about...
Sodium

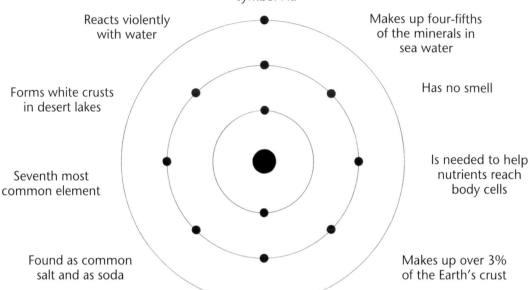

A silvery-coloured metal, chemical symbol Na

Reacts violently with water

Makes up four-fifths of the minerals in sea water

Forms white crusts in desert lakes

Has no smell

Seventh most common element

Is needed to help nutrients reach body cells

Found as common salt and as soda

Makes up over 3% of the Earth's crust

Atomic number 11, atomic weight about 23

SHELL DIAGRAMS

The shell diagrams on these pages are representations of an atom of each element. The total number of electrons are shown in the relevant orbitals, or shells, around the central nucleus.

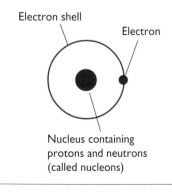

Electron shell

Electron

Nucleus containing protons and neutrons (called nucleons)

▶ Sodium makes up four-fifths of the minerals in sea water.

Potassium

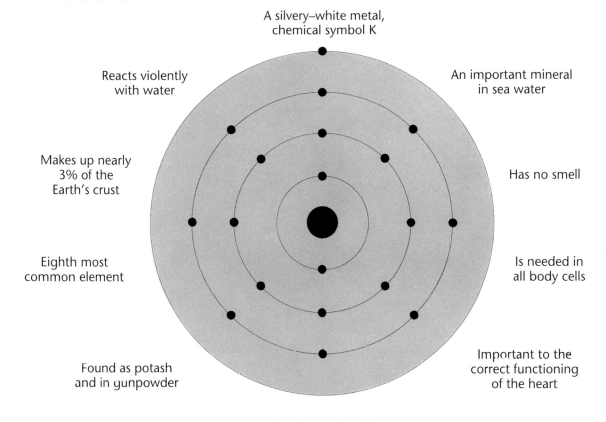

A silvery–white metal, chemical symbol K

Reacts violently with water

An important mineral in sea water

Makes up nearly 3% of the Earth's crust

Has no smell

Eighth most common element

Is needed in all body cells

Found as potash and in gunpowder

Important to the correct functioning of the heart

Atomic number 19, atomic weight about 39

▼ Sodium carbonate is used in toothpaste to neutralise acids that corrode teeth.

The Periodic Table

The Periodic Table sets out the relationships among the elements of the Universe. According to the Periodic Table, certain elements fall into groups. The pattern of these groups has, in the past, allowed scientists to predict elements that had not at that time been discovered. It can still be used today to predict the properties of unfamiliar elements.

The Periodic Table was first described by a Russian teacher, Dmitry Ivanovich Mendeleev, between 1869 and 1870. He was interested in writing a chemistry textbook, and wanted to show his students that there were certain patterns in the elements that had been discovered. So he set out the elements (of which there were 57 at the time) according to their known properties. On the assumption that there was pattern to the elements, he left blank spaces where elements seemed to be missing. Using this first version of the Periodic Table, he was able to predict in detail the chemical and physical properties of elements that had not yet been discovered. Other scientists began to look for the missing elements, and they soon found them.

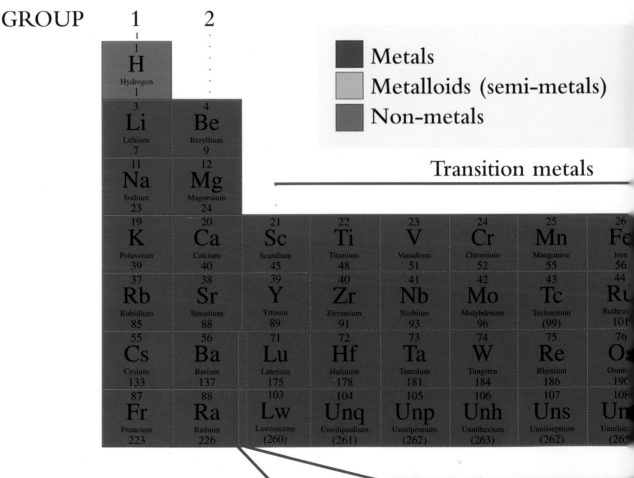

GROUP

Metals
Metalloids (semi-metals)
Non-metals

Transition metals

Lanthanide metals

Actinoid metals

Hydrogen did not seem to fit into the table, so he placed it in a box on its own. Otherwise the elements were all placed horizontally. When an element was reached with properties similar to the first one in the top row, a second row was started. By following this rule, similarities among the elements can be found by reading up and down. By reading across the rows, the elements progressively increase their atomic number. This number indicates the number of positively charged particles (protons) in the nucleus of each atom. This is also the number of negatively charged particles (electrons) in the atom.

The chemical properties of an element depend on the number of electrons in the outermost shell.

Atoms can form compounds by sharing electrons in their outermost shells. This explains why atoms with a full set of electrons (like helium, an inert gas) are unreactive, whereas atoms with an incomplete electron shell (such as chlorine) are very reactive. Elements can also combine by the complete transfer of electrons from metals to non-metals and the compounds formed contain ions.

Radioactive elements lose particles from their nucleus and electrons from their surrounding shells. As a result their atomic number changes and they become new elements.

Atomic (proton) number
13 — Symbol
Al
Aluminium — Name
27
Approximate relative atomic mass
(Approximate atomic weight)

3	4	5	6	7	0
					2 He Helium 4
5 B Boron 11	6 C Carbon 12	7 N Nitrogen 14	8 O Oxygen 16	9 F Fluorine 19	10 Ne Neon 20
13 Al Aluminium 27	14 Si Silicon 28	15 P Phosphorus 31	16 S Sulphur 32	17 Cl Chlorine 35	18 Ar Argon 40

27 Co Cobalt 59	28 Ni Nickel 59	29 Cu Copper 64	30 Zn Zinc 65	31 Ga Gallium 70	32 Ge Germanium 73	33 As Arsenic 75	34 Se Selenium 79	35 Br Bromine 80	36 Kr Krypton 84
45 Rh Rhodium 103	46 Pd Palladium 106	47 Ag Silver 108	48 Cd Cadmium 112	49 In Indium 115	50 Sn Tin 119	51 Sb Antimony 122	52 Te Tellurium 128	53 I Iodine 127	54 Xe Xenon 131
77 Ir Iridium 192	78 Pt Platinum 195	79 Au Gold 197	80 Hg Mercury 201	81 Tl Thallium 204	82 Pb Lead 207	83 Bi Bismuth 209	84 Po Polonium (209)	85 At Astatine (210)	86 Rn Radon (222)
109 Une Unnilennium (266)									

61 Pm Promethium (145)	62 Sm Samarium 150	63 Eu Europium 152	64 Gd Gadolinium 157	65 Tb Terbium 159	66 Dy Dysprosium 163	67 Ho Holmium 165	68 Er Erbium 167	69 Tm Thulium 169	70 Yb Ytterbium 173
93 Np Neptunium (237)	94 Pu Plutonium (244)	95 Am Americium (243)	96 Cm Curium (247)	97 Bk Berkelium (247)	98 Cf Californium (251)	99 Es Einsteinium (252)	100 Fm Fermium (257)	101 Md Mendelevium (258)	102 No Nobelium (259)

Understanding equations

As you read through this book, you will notice that many pages contain equations using symbols. If you are not familiar with these symbols, read this page. Symbols make it easy for chemists to write out the reactions that are occurring in a way that allows a better understanding of the processes involved.

Symbols for the elements

The basis of the modern use of symbols for elements dates back to the 19th century. At this time a shorthand was developed using the first letter of the element wherever possible. Thus "O" stands for oxygen, "H" stands for hydrogen and so on. However, if we were to use only the first letter, then there could be some confusion. For example, nitrogen and nickel would both use the symbols N. To overcome this problem, many elements are symbolised using the first two letters of their full name, and the second letter is lowercase. Thus although nitrogen is N, nickel becomes Ni. Not all symbols come from the English name; many use the Latin name instead. This is why, for example, gold is not G but Au (for the Latin *aurum*) and sodium has the symbol Na, from the Latin *natrium*.

Compounds of elements are made by combining letters. Thus the molecule carbon

Written and symbolic equations
In this book, important chemical equations are briefly stated in words (these are called word equations), and are then shown in their symbolic form along with the states.

What reaction the equation illustrates

EQUATION: The formation of calcium hydroxide

Word equation —————— *Calcium oxide + water ⇨ calcium hydroxide*

Symbol equation —————— $CaO(s) \quad + \quad H_2O(l) \quad ⇨ \quad Ca(OH)_2(aq)$

heated

Sometimes you will find additional descriptions below the symbolic equation.

Symbol showing the state:
s is for solid, l is for liquid, g is for gas and aq is for aqueous.

Diagrams
Some of the equations are shown as graphic representations.

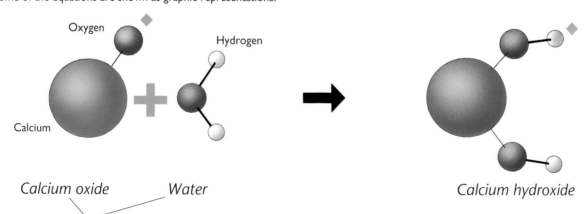

Oxygen

Hydrogen

Calcium

Calcium oxide   *Water*

Calcium hydroxide

Sometimes the written equation is broken up and put below the relevant stages in the graphic representation.

monoxide is CO. By using lowercase letters for the second letter of an element, it is possible to show that cobalt, symbol Co, is not the same as the molecule carbon monoxide, CO.

However, the letters can be made to do much more than this. In many molecules, atoms combine in unequal numbers. So, for example, carbon dioxide has one atom of carbon for every two of oxygen. This is shown by using the number 2 beside the oxygen, and the symbol becomes CO_2.

In practice, some groups of atoms combine as a unit with other substances. Thus, for example, calcium bicarbonate (one of the compounds used in some antacid pills) is written $Ca(HCO_3)_2$. This shows that the part of the substance inside the brackets reacts as a unit and the "2" outside the brackets shows the presence of two such units.

Some substances attract water molecules to themselves. To show this a dot is used. Thus the blue form of copper sulphate is written $CuSO_4.5H_2O$. In this case five molecules of water attract to one of copper sulphate.

When you see the dot, you know that this water can be driven off by heating; it is part of the crystal structure.

In a reaction substances change by rearranging the combinations of atoms. The way they change is shown by using the chemical symbols, placing those that will react (the starting materials, or reactants) on the left and the products of the reaction on the right. Between the two, chemists use an arrow to show which way the reaction is occurring.

It is possible to describe a reaction in words. This gives word equations, which are given throughout this book. However, it is easier to understand what is happening by using an equation containing symbols. These are also given in many places. They are not given when the equations are very complex.

In any equation both sides balance; that is, there must be an equal number of like atoms on both sides of the arrow. When you try to write down reactions, you, too, must balance your equation; you cannot have a few atoms left over at the end!

The symbols in brackets are abbreviations for the physical state of each substance taking part, so that (s) is used for solid, (l) for liquid, (g) for gas and (aq) for an aqueous solution, that is, a solution of a substance dissolved in water.

Atoms and ions
Each sphere represents a particle of an element. A particle can be an atom or an ion. Each atom or ion is associated with other atoms or ions through bonds – forces of attraction. The size of the particles and the nature of the bonds can be extremely important in determining the nature of the reaction or the properties of the compound.

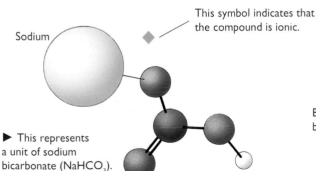

Sodium

This symbol indicates that the compound is ionic.

▶ This represents a unit of sodium bicarbonate ($NaHCO_3$).

The term "unit" is sometimes used to simplify the representation of a combination of ions.

Chemical symbols, equations and diagrams
The arrangement of any molecule or compound can be shown in one of the two ways shown below, depending on which gives the clearer picture. The left-hand diagram is called a ball-and-stick diagram because it uses rods and spheres to show the structure of the material. This example shows water, H_2O. There are two hydrogen atoms and one oxygen atom.

Bond shown by "stick"

Colours too
The colours of each of the particles help differentiate the elements involved. The diagram can then be matched to the written and symbolic equation given with the diagram. In the case above, oxygen is red and hydrogen is grey.

Glossary of technical terms

absorb: to soak up a substance. Compare to adsorb.

acetone: a petroleum-based solvent.

acid: compounds containing hydrogen which can attack and dissolve many substances. Acids are described as weak or strong, dilute or concentrated, mineral or organic.

acidity: a general term for the strength of an acid in a solution.

acid rain: rain that is contaminated by acid gases such as sulphur dioxide and nitrogen oxides released by pollution.

adsorb/adsorption: to "collect" gas molecules or other particles on to the *surface* of a substance. They are not chemically combined and can be removed. (The process is called "adsorption".) Compare to absorb.

alchemy: the traditional "art" of working with chemicals that prevailed through the Middle Ages. One of the main challenges of alchemy was to make gold from lead. Alchemy faded away as scientific chemistry was developed in the 17th century.

alkali: a base in solution.

alkaline: the opposite of acidic. Alkalis are bases that dissolve, and alkaline materials are called basic materials. Solutions of alkalis have a pH greater than 7.0 because they contain relatively few hydrogen ions.

alloy: a mixture of a metal and various other elements.

alpha particle: a stable combination of two protons and two neutrons, which is ejected from the nucleus of a radioactive atom as it decays. An alpha particle is also the nucleus of the atom of helium. If it captures two electrons it can become a neutral helium atom.

amalgam: a liquid alloy of mercury with another metal.

amino acid: amino acids are organic compounds that are the building blocks for the proteins in the body.

amorphous: a solid in which the atoms are not arranged regularly (i.e. "glassy"). Compare with crystalline.

amphoteric: a metal that will react with both acids and alkalis.

anhydrous: a substance from which water has been removed by heating. Many hydrated salts are crystalline. When they are heated and the water is driven off, the material changes to an anhydrous powder.

anion: a negatively charged atom or group of atoms.

anode: the negative terminal of a battery or the positive electrode of an electrolysis cell.

anodising: a process that uses the effect of electrolysis to make a surface corrosion-resistant.

antacid: a common name for any compound that reacts with stomach acid to neutralise it.

antioxidant: a substance that prevents oxidation of some other substance.

aqueous: a solid dissolved in water. Usually used as "aqueous solution".

atom: the smallest particle of an element.

atomic number: the number of electrons or the number of protons in an atom.

atomised: broken up into a very fine mist. The term is used in connection with sprays and engine fuel systems.

aurora: the "northern lights" and "southern lights" that show as coloured bands of light in the night sky at high latitudes. They are associated with the way cosmic rays interact with oxygen and nitrogen in the air.

basalt: an igneous rock with a low proportion of silica (usually below 55%). It has microscopically small crystals.

base: a compound that may be soapy to the touch and that can react with an acid in water to form a salt and water.

battery: a series of electrochemical cells.

bauxite: an ore of aluminium, of which about half is aluminium oxide.

becquerel: a unit of radiation equal to one nuclear disintegration per second.

beta particle: a form of radiation in which electrons are emitted from an atom as the nucleus breaks down.

bleach: a substance that removes stains from materials either by oxidising or reducing the staining compound.

boiling point: the temperature at which a liquid boils, changing from a liquid to a gas.

bond: chemical bonding is either a transfer or sharing of electrons by two or more atoms. There are a number of types of chemical bond, some very strong (such as covalent bonds), others weak (such as hydrogen bonds). Chemical bonds form because the linked molecule is more stable than the unlinked atoms from which it formed. For example, the hydrogen molecule (H_2) is more stable than single atoms of hydrogen, which is why hydrogen gas is always found as molecules of two hydrogen atoms.

brass: a metal alloy principally of copper and zinc.

brazing: a form of soldering, in which brass is used as the joining metal.

brine: a solution of salt (sodium chloride) in water.

bronze: an alloy principally of copper and tin.

buffer: a chemistry term meaning a mixture of substances in solution that resists a change in the acidity or alkalinity of the solution.

capillary action: the tendency of a liquid to be sucked into small spaces, such as between objects and through narrow-pore tubes. The force to do this comes from surface tension.

catalyst: a substance that speeds up a chemical reaction but itself remains unaltered at the end of the reaction.

cathode: the positive terminal of a battery or the negative electrode of an electrolysis cell.

cathodic protection: the technique of making the object that is to be protected from corrosion into the cathode of a cell. For example, a material, such as steel, is protected by coupling it with a more reactive metal, such as magnesium. Steel forms the cathode and magnesium the anode. Zinc protects steel in the same way.

cation: a positively charged atom or group of atoms.

caustic: a substance that can cause burns if it touches the skin.

cell: a vessel containing two electrodes and an electrolyte that can act as an electrical conductor.

ceramic: a material based on clay minerals, which has been heated so that it has chemically hardened.

chalk: a pure form of calcium carbonate made of the crushed bodies of microscopic sea creatures, such as plankton and algae.

change of state: a change between one of the three states of matter, solid, liquid and gas.

chlorination: adding chlorine to a substance.

cladding: a surface sheet of material designed to protect other materials from corrosion.

clay: a microscopically small plate-like mineral that makes up the bulk of many soils. It has a sticky feel when wet.

combustion: the special case of oxidisation of a substance where a considerable amount of heat and usually light are given out. Combustion is often referred to as "burning".

compound: a chemical consisting of two or more elements chemically bonded together. Calcium atoms can combine with carbon atoms and oxygen atoms to make calcium carbonate, a compound of all three atoms.

condensation nuclei: microscopic particles of dust, salt and other materials suspended in the air, which attract water molecules.

conduction: (i) the exchange of heat (heat conduction) by contact with another object or (ii) allowing the flow of electrons (electrical conduction).

convection: the exchange of heat energy with the surroundings produced by the flow of a fluid due to being heated or cooled.

corrosion: the *slow* decay of a substance resulting from contact with gases and liquids in the environment. The term is often applied to metals. Rust is the corrosion of iron.

corrosive: a substance, either an acid or an alkali, that *rapidly* attacks a wide range of other substances.

cosmic rays: particles that fly through space and bombard all atoms on the Earth's surface. When they interact with the atmosphere they produce showers of secondary particles.

covalent bond: the most common form of strong chemical bonding, which occurs when two atoms *share* electrons.

cracking: breaking down complex molecules into simpler components. It is a term particularly used in oil refining.

crude oil: a chemical mixture of petroleum liquids. Crude oil forms the raw material for an oil refinery.

crystal: a substance that has grown freely so that it can develop external faces. Compare with crystalline, where the atoms are not free to form individual crystals and amorphous where the atoms are arranged irregularly.

crystalline: the organisation of atoms into a rigid "honeycomb-like" pattern without distinct crystal faces.

crystal systems: seven patterns or systems into which all of the world's crystals can be grouped. They are: cubic, hexagonal, rhombohedral, tetragonal, orthorhombic, monoclinic and triclinic.

cubic crystal system: groupings of crystals that look like cubes.

curie: a unit of radiation. The amount of radiation emitted by 1 g of radium each second. (The curie is equal to 37 billion becquerels.)

current: an electric current is produced by a flow of electrons through a conducting solid or ions through a conducting liquid.

decay (radioactive decay): the way that a radioactive element changes into another element because of loss of mass through radiation. For example uranium decays (changes) to lead.

decompose: to break down a substance (for example by heat or with the aid of a catalyst) into simpler components. In such a chemical reaction only one substance is involved.

dehydration: the removal of water from a substance by heating it, placing it in a dry atmosphere, or through the action of a drying agent.

density: the mass per unit volume (e.g. g/cc).

desertification: a process whereby a soil is allowed to become degraded to a state in which crops can no longer grow, i.e. desert-like. Chemical desertification is usually the result of contamination with halides because of poor irrigation practices.

detergent: a petroleum-based chemical that removes dirt.

diaphragm: a semipermeable membrane – a kind of ultra-fine mesh filter – that will allow only small ions to pass through. It is used in the electrolysis of brine.

diffusion: the slow mixing of one substance with another until the two substances are evenly mixed.

digestive tract: the system of the body that forms the pathway for food and its waste products. It begins at the mouth and includes the stomach and the intestines.

dilute acid: an acid whose concentration has been reduced by a large proportion of water.

diode: a semiconducting device that allows an electric current to flow in only one direction.

disinfectant: a chemical that kills bacteria and other microorganisms.

dissociate: to break apart. In the case of acids it means to break up forming hydrogen ions. This is an example of ionisation. Strong acids dissociate completely. Weak acids are not completely ionised and a solution of a weak acid has a relatively low concentration of hydrogen ions.

dissolve: to break down a substance in a solution without a resultant reaction.

distillation: the process of separating mixtures by condensing the vapours through cooling.

doping: adding metal atoms to a region of silicon to make it semiconducting.

dye: a coloured substance that will stick to another substance, so that both appear coloured.

electrode: a conductor that forms one terminal of a cell.

electrolysis: an electrical–chemical process that uses an electric current to cause the break up of a compound and the movement of metal ions in a solution. The process happens in many natural situations (as for example in rusting) and is also commonly used in industry for purifying (refining) metals or for plating metal objects with a fine, even metal coating.

electrolyte: a solution that conducts electricity.

electron: a tiny, negatively charged particle that is part of an atom. The flow of electrons through a solid material such as a wire produces an electric current.

electroplating: depositing a thin layer of a metal onto the surface of another substance using electrolysis.

element: a substance that cannot be decomposed into simpler substances by chemical means

emulsion: tiny droplets of one substance dispersed in another. A common oil in water emulsion is milk. The tiny droplets in an emulsion tend to come together, so another stabilising substance is often needed to wrap the particles of grease and oil in a stable coat. Soaps and detergents are such agents. Photographic film is an example of a solid emulsion.

endothermic reaction: a reaction that takes heat from the surroundings. The reaction of carbon monoxide with a metal oxide is an example.

enzyme: organic catalysts in the form of proteins in the body that speed up chemical reactions. Every living cell contains hundreds of enzymes, which ensure that the processes of life continue. Should enzymes be made inoperative, such as through mercury poisoning, then death follows.

ester: organic compounds, formed by the reaction of an alcohol with an acid, which often have a fruity taste.

evaporation: the change of state of a liquid to a gas. Evaporation happens below the boiling point and is used as a method of separating out the materials in a solution.

exothermic reaction: a reaction that gives heat to the surroundings. Many oxidation reactions, for example, give out heat.

explosive: a substance which, when a shock is applied to it, decomposes very rapidly, releasing a very large amount of heat and creating a large volume of gases as a shock wave.

extrusion: forming a shape by pushing it through a die. For example, toothpaste is extruded through the cap (die) of the toothpaste tube.

fallout: radioactive particles that reach the ground from radioactive materials in the atmosphere.

fat: semi-solid energy-rich compounds derived from plants or animals and which are made of carbon, hydrogen and oxygen. Scientists call these esters.

feldspar: a mineral consisting of sheets of aluminium silicate. This is the mineral from which the clay in soils is made.

fertile: able to provide the nutrients needed for unrestricted plant growth.

filtration: the separation of a liquid from a solid using a membrane with small holes.

fission: the breakdown of the structure of an atom, popularly called "splitting the atom" because the atom is split into approximately two other nuclei. This is different from, for example, the small change that happens when radioactivity is emitted.

fixation of nitrogen: the processes that natural organisms, such as bacteria, use to turn the nitrogen of the air into ammonium compounds.

fixing: making solid and liquid nitrogen-containing compounds from nitrogen gas. The compounds that are formed can be used as fertilisers.

fluid: able to flow; either a liquid or a gas.

fluorescent: a substance that gives out visible light when struck by invisible waves such as ultraviolet rays.

flux: a material used to make it easier for a liquid to flow. A flux dissolves metal oxides and so prevents a metal from oxidising while being heated.

foam: a substance that is sufficiently gelatinous to be able to contain bubbles of gas. The gas bulks up the substance, making it behave as though it were semi-rigid.

fossil fuels: hydrocarbon compounds that have been formed from buried plant and animal remains. High pressures and temperatures lasting over millions of years are required. The fossil fuels are coal, oil and natural gas.

fraction: a group of similar components of a mixture. In the petroleum industry the light fractions of crude oil are those with the smallest molecules, while the medium and heavy fractions have larger molecules.

free radical: a very reactive atom or group with a "spare" electron.

freezing point: the temperature at which a substance changes from a liquid to a solid. It is the same temperature as the melting point.

fuel: a concentrated form of chemical energy. The main sources of fuels (called fossil fuels because they were formed by geological processes) are coal, crude oil and natural gas. Products include methane, propane and gasoline. The fuel for stars and space vehicles is hydrogen.

fuel rods: rods of uranium or other radioactive material used as a fuel in nuclear power stations.

fuming: an unstable liquid that gives off a gas. Very concentrated acid solutions are often fuming solutions.

fungicide: any chemical that is designed to kill fungi and control the spread of fungal spores.

fusion: combining atoms to form a heavier atom.

galvanising: applying a thin zinc coating to protect another metal.

gamma rays: waves of radiation produced as the nucleus of a radioactive element rearranges itself into a tighter cluster of protons and neutrons. Gamma rays carry enough energy to damage living cells.

gangue: the unwanted material in an ore.

gas: a form of matter in which the molecules form no definite shape and are free to move about to fill any vessel they are put in.

gelatinous: a term meaning made with water. Because a gelatinous precipitate is mostly water, it is of a similar density to water and will float or lie suspended in the liquid.

gelling agent: a semi-solid jelly-like substance.

gemstone: a wide range of minerals valued by people, both as crystals (such as emerald) and as decorative stones (such as agate). There is no single chemical formula for a gemstone.

glass: a transparent silicate without any crystal growth. It has a glassy lustre and breaks with a curved fracture. Note that some minerals have all these features and are therefore natural glasses. Household glass is a synthetic silicate.

glucose: the most common of the natural sugars. It occurs as the polymer known as cellulose, the fibre in plants. Starch is also a form of glucose. The breakdown of glucose provides the energy that animals need for life.

granite: an igneous rock with a high proportion of silica (usually over 65%). It has well-developed large crystals. The largest pink, grey or white crystals are feldspar.

Greenhouse Effect: an increase of the global air temperature as a result of heat released from burning fossil fuels being absorbed by carbon dioxide in the atmosphere.

gypsum: the name for calcium sulphate. It is commonly found as Plaster of Paris and wallboards.

half-life: the time it takes for the radiation coming from a sample of a radioactive element to decrease by half.

halide: a salt of one of the halogens (fluorine, chlorine, bromine and iodine).

halite: the mineral made of sodium chloride.

halogen: one of a group of elements including chlorine, bromine, iodine and fluorine.

heat-producing: see exothermic reaction.

high explosive: a form of explosive that will only work when it receives a shock from another explosive. High explosives are much more powerful than ordinary explosives. Gunpowder is not a high explosive.

hydrate: a solid compound in crystalline form that contains molecular water. Hydrates commonly form when a solution of a soluble salt is evaporated. The water that forms part of a hydrate crystal is known as the "water of crystallization". It can usually be removed by heating, leaving an anhydrous salt.

hydration: the absorption of water by a substance. Hydrated materials are not "wet" but remain firm, apparently dry, solids. In some cases, hydration makes the substance change colour, in many other cases there is no colour change, simply a change in volume.

hydrocarbon: a compound in which only hydrogen and carbon atoms are present. Most fuels are hydrocarbons, as is the simple plastic polyethene (known as polythene).

hydrogen bond: a type of attractive force that holds one molecule to another. It is one of the weaker forms of intermolecular attractive force.

hydrothermal: a process in which hot water is involved. It is usually used in the context of rock formation because hot water and other fluids sent outwards from liquid magmas are important carriers of metals and the minerals that form gemstones.

igneous rock: a rock that has solidified from molten rock, either volcanic lava on the Earth's surface or magma deep underground. In either case the rock develops a network of interlocking crystals.

incendiary: a substance designed to cause burning.

indicator: a substance or mixture of substances that change colour with acidity or alkalinity.

inert: nonreactive.

infra-red radiation: a form of light radiation where the wavelength of the waves is slightly longer than visible light. Most heat radiation is in the infra-red band.

insoluble: a substance that will not dissolve.

ion: an atom, or group of atoms, that has gained or lost one or more electrons and so developed an electrical charge. Ions behave differently from electrically neutral atoms and molecules. They can move in an electric field, and they can also bind strongly to solvent molecules such as water. Positively charged ions are called cations; negatively charged ions are called anions. Ions carry electrical current through solutions.

ionic bond: the form of bonding that occurs between two ions when the ions have opposite charges. Sodium cations bond with chloride anions to form common salt (NaCl) when a salty solution is evaporated. Ionic bonds are strong bonds except in the presence of a solvent.

ionise: to break up neutral molecules into oppositely charged ions or to convert atoms into ions by the loss of electrons.

ionisation: a process that creates ions.

irrigation: the application of water to fields to help plants grow during times when natural rainfall is sparse.

isotope: atoms that have the same number of protons in their nucleus, but which have different masses; for example, carbon-12 and carbon-14.

latent heat: the amount of heat that is absorbed or released during the process of changing state between gas, liquid or solid. For example, heat is absorbed when a substance melts and it is released again when the substance solidifies.

latex. (the Latin word for "liquid") a suspension of small polymer particles in water. The rubber that flows from a rubber tree is a natural latex. Some synthetic polymers are made as latexes, allowing polymerisation to take place in water.

lava: the material that flows from a volcano.

limestone: a form of calcium carbonate rock that is often formed of lime mud. Most limestones are light grey and have abundant fossils.

liquid: a form of matter that has a fixed volume but no fixed shape.

lode: a deposit in which a number of veins of a metal found close together.

lustre: the shininess of a substance.

magma: the molten rock that forms a balloon-shaped chamber in the rock below a volcano. It is fed by rock moving upwards from below the crust.

marble: a form of limestone that has been "baked" while deep inside mountains. This has caused the limestone to melt and reform into small interlocking crystals, making marble harder than limestone.

mass: the amount of matter in an object. In everyday use, the word weight is often used to mean mass.

melting point: the temperature at which a substance changes state from a solid to a liquid. It is the same as freezing point.

membrane: a thin flexible sheet. A semipermeable membrane has microscopic holes of a size that will selectively allow some ions and molecules to pass through but hold others back. It thus acts as a kind of sieve.

meniscus: the curved surface of a liquid that forms when it rises in a small bore, or capillary tube. The meniscus is convex (bulges upwards) for mercury and is concave (sags downwards) for water.

metal: a substance with a lustre, the ability to conduct heat and electricity and which is not brittle.

metallic bonding: a kind of bonding in which atoms reside in a "sea" of mobile electrons. This type of bonding allows metals to be good conductors and means that they are not brittle

metamorphic rock: formed either from igneous or sedimentary rocks, by heat and or pressure. Metamorphic rocks form deep inside mountains during periods of mountain building. They result from the remelting of rocks during which process crystals are able to grow. Metamorphic rocks often show signs of banding and partial melting.

micronutrient: an element that the body requires in small amounts. Another term is trace element.

mineral: a solid substance made of just one element or chemical compound. Calcite is a mineral because it consists only of calcium carbonate, halite is a mineral because it contains only sodium chloride, quartz is a mineral because it consists of only silicon dioxide.

mineral acid: an acid that does not contain carbon and that attacks minerals. Hydrochloric, sulphuric and nitric acids are the main mineral acids.

mineral-laden: a solution close to saturation.

mixture: a material that can be separated out into two or more substances using physical means.

molecule: a group of two or more atoms held together by chemical bonds.

monoclinic system: a grouping of crystals that look like double-ended chisel blades.

monomer: a building block of a larger chain molecule ("mono" means one, "mer" means part).

mordant: any chemical that allows dyes to stick to other substances.

native metal: a pure form of a metal, not combined as a compound. Native metal is more common in poorly reactive elements than in those that are very reactive.

neutralisation: the reaction of acids and bases to produce a salt and water. The reaction causes hydrogen from the acid and hydroxide from the base to be changed to water. For example, hydrochloric acid reacts with sodium hydroxide to form common salt and water. The term is more generally used for any reaction where the pH changes towards 7.0, which is the pH of a neutral solution.

neutron: a particle inside the nucleus of an atom that is neutral and has no charge.

noncombustible: a substance that will not burn.

noble metal: silver, gold, platinum, and mercury. These are the least reactive metals.

nuclear energy: the heat energy produced as part of the changes that take place in the core, or nucleus, of an element's atoms.

nuclear reactions: reactions that occur in the core, or nucleus of an atom.

nutrients: soluble ions that are essential to life.

octane: one of the substances contained in fuel.

ore: a rock containing enough of a useful substance to make mining it worthwhile.

organic acid: an acid containing carbon and hydrogen.

organic substance: a substance that contains carbon.

osmosis: a process where molecules of a liquid solvent move through a membrane (filter) from a region of low concentration to a region of high concentration of solute.

oxidation: a reaction in which the oxidising agent removes electrons. (Note that oxidising agents do not have to contain oxygen.)

oxide: a compound that includes oxygen and one other element.

oxidise: the process of gaining oxygen. This can be part of a controlled chemical reaction, or it can be the result of exposing a substance to the air, where oxidation (a form of corrosion) will occur slowly, perhaps over months or years.

oxidising agent: a substance that removes electrons from another substance (and therefore is itself reduced).

ozone: a form of oxygen whose molecules contain three atoms of oxygen. Ozone is regarded as a beneficial gas when high in the atmosphere because it blocks ultraviolet rays. It is a harmful gas when breathed in, so low level ozone, which is produced as part of city smog, is regarded as a form of pollution. The ozone layer is the uppermost part of the stratosphere.

pan: the name given to a shallow pond of liquid. Pans are mainly used for separating solutions by evaporation.

patina: a surface coating that develops on metals and protects them from further corrosion.

percolate: to move slowly through the pores of a rock.

period: a row in the Periodic Table.

Periodic Table: a chart organising elements by atomic number and chemical properties into groups and periods.

pesticide: any chemical that is designed to control pests (unwanted organisms) that are harmful to plants or animals.

petroleum: a natural mixture of a range of gases, liquids and solids derived from the decomposed remains of plants and animals.

pH: a measure of the hydrogen ion concentration in a liquid. Neutral is pH 7.0; numbers greater than this are alkaline, smaller numbers are acidic.

phosphor: any material that glows when energized by ultraviolet or electron beams such as in fluorescent tubes and cathode ray tubes. Phosphors, such as phosphorus, emit light after the source of excitation is cut off. This is why they glow in the dark. By contrast, fluorescors, such as fluorite, emit light only while they are being excited by ultraviolet light or an electron beam.

photon: a parcel of light energy.

photosynthesis: the process by which plants use the energy of the Sun to make the compounds they need for life. In photosynthesis, six molecules of carbon dioxide from the air combine with six molecules of water, forming one molecule of glucose (sugar) and releasing six molecules of oxygen back into the atmosphere.

pigment: any solid material used to give a liquid a colour.

placer deposit: a kind of ore body made of a sediment that contains fragments of gold ore eroded from a mother lode and transported by rivers and/or ocean currents.

plastic (material): a carbon-based material consisting of long chains (polymers) of simple molecules. The word plastic is commonly restricted to synthetic polymers.

plastic (property): a material is plastic if it can be made to change shape easily. Plastic materials will remain in the new shape. (Compare with elastic, a property where a material goes back to its original shape.)

plating: adding a thin coat of one material to another to make it resistant to corrosion.

playa: a dried-up lake bed that is covered with salt deposits. From the Spanish word for beach.

poison gas: a form of gas that is used intentionally to produce widespread injury and death. (Many gases are poisonous, which is why many chemical reactions are performed in laboratory fume chambers, but they are a byproduct of a reaction and not intended to cause harm.)

polymer: a compound that is made of long chains by combining molecules (called monomers) as repeating units. ("Poly" means many, "mer" means part).

polymerisation: a chemical reaction in which large numbers of similar molecules arrange themselves into large molecules, usually long chains. This process usually happens when there is a suitable catalyst present. For example, ethene reacts to form polythene in the presence of certain catalysts.

porous: a material containing many small holes or cracks. Quite often the pores are connected, and liquids, such as water or oil, can move through them.

precious metal: silver, gold, platinum, iridium, and palladium. Each is prized for its rarity. This category is the equivalent of precious stones, or gemstones, for minerals.

precipitate: tiny solid particles formed as a result of a chemical reaction between two liquids or gases.

preservative: a substance that prevents the natural organic decay processes from occurring. Many substances can be used safely for this purpose, including sulphites and nitrogen gas.

product: a substance produced by a chemical reaction.

protein: molecules that help to build tissue and bone and therefore make new body cells. Proteins contain amino acids.

proton: a positively charged particle in the nucleus of an atom that balances out the charge of the surrounding electrons

pyrite: "mineral of fire". This name comes from the fact that pyrite (iron sulphide) will give off sparks if struck with a stone.

pyrometallurgy: refining a metal from its ore using heat. A blast furnace or smelter is the main equipment used.

radiation: the exchange of energy with the surroundings through the transmission of waves or particles of energy. Radiation is a form of energy transfer that can happen through space; no intervening medium is required (as would be the case for conduction and convection).

radioactive: a material that emits radiation or particles from the nucleus of its atoms.

radioactive decay: a change in a radioactive element due to loss of mass through radiation. For example uranium decays (changes) to lead.

radioisotope: a shortened version of the phrase radioactive isotope.

radiotracer: a radioactive isotope that is added to a stable, nonradioactive material in order to trace how it moves and its concentration.

reaction: the recombination of two substances using parts of each substance to produce new substances.

reactivity: the tendency of a substance to react with other substances. The term is most widely used in comparing the reactivity of metals. Metals are arranged in a reactivity series.

reagent: a starting material for a reaction.

recycling: the reuse of a material to save the time and energy required to extract new material from the Earth and to conserve non-renewable resources.

redox reaction: a reaction that involves reduction and oxidation.

reducing agent: a substance that gives electrons to another substance. Carbon monoxide is a reducing agent when passed over copper oxide, turning it to copper and producing carbon dioxide gas. Similarly, iron oxide is reduced to iron in a blast furnace. Sulphur dioxide is a reducing agent, used for bleaching bread.

reduction: the removal of oxygen from a substance. See also: oxidation.

refining: separating a mixture into the simpler substances of which it is made. In the case of a rock, it means the extraction of the metal that is mixed up in the rock. In the case of oil it means separating out the fractions of which it is made.

refractive index: the property of a transparent material that controls the angle at which total internal reflection will occur. The greater the refractive index, the more reflective the material will be.

resin: natural or synthetic polymers that can be moulded into solid objects or spun into thread.

rust: the corrosion of iron and steel.

saline: a solution in which most of the dissolved matter is sodium chloride (common salt).

salinisation: the concentration of salts, especially sodium chloride, in the upper layers of a soil due to poor methods of irrigation.

salts: compounds, often involving a metal, that are the reaction products of acids and bases. (Note "salt" is also the common word for sodium chloride, common salt or table salt.)

saponification: the term for a reaction between a fat and a base that produces a soap.

saturated: a state where a liquid can hold no more of a substance. If any more of the substance is added, it will not dissolve.

saturated solution: a solution that holds the maximum possible amount of dissolved material. The amount of material in solution varies with the temperature; cold solutions can hold less dissolved solid material than hot solutions. Gases are more soluble in cold liquids than hot liquids.

sediment: material that settles out at the bottom of a liquid when it is still.

semiconductor: a material of intermediate conductivity. Semiconductor devices often use silicon when they are made as part of diodes, transistors or integrated circuits.

semipermeable membrane: a thin (membrane) of material that acts as a fine sieve, allowing small molecules to pass, but holding large molecules back.

silicate: a compound containing silicon and oxygen (known as silica).

sintering: a process that happens at moderately high temperatures in some compounds. Grains begin to fuse together even through they do not melt. The most widespread example of sintering happens during the firing of clays to make ceramics.

slag: a mixture of substances that are waste products of a furnace. Most slags are composed mainly of silicates.

smelting: roasting a substance in order to extract the metal contained in it.

smog: a mixture of smoke and fog. The term is used to describe city fogs in which there is a large proportion of particulate matter (tiny pieces of carbon from exhausts) and also a high concentration of sulphur and nitrogen gases and probably ozone.

soldering: joining together two pieces of metal using solder, an alloy with a low melting point.

solid: a form of matter where a substance has a definite shape.

soluble: a substance that will readily dissolve in a solvent.

solute: the substance that dissolves in a solution (e.g. sodium chloride in salt water).

solution: a mixture of a liquid and at least one other substance (e.g. salt water). Mixtures can be separated out by physical means, for example by evaporation and cooling.

solvent: the main substance in a solution (e.g. water in salt water).

spontaneous combustion: the effect of a very reactive material beginning to oxidise very quickly and bursting into flame.

stable: able to exist without changing into another substance.

stratosphere: the part of the Earth's atmosphere that lies immediately above the region in which clouds form. It occurs between 12 and 50 km above the Earth's surface.

strong acid: an acid that has completely dissociated (ionised) in water. Mineral acids are strong acids.

sublimation: the change of a substance from solid to gas, or vica versa, without going through a liquid phase.

substance: a type of material, including mixtures.

sulphate: a compound that includes sulphur and oxygen, for example, calcium sulphate or gypsum.

sulphide: a sulphur compound that contains no oxygen.

sulphite: a sulphur compound that contains less oxygen than a sulphate.

surface tension: the force that operates on the surface of a liquid, which makes it act as though it were covered with an invisible elastic film.

suspension: tiny particles suspended in a liquid.

synthetic: does not occur naturally, but has to be manufactured.

tarnish: a coating that develops as a result of the reaction between a metal and substances in the air. The most common form of tarnishing is a very thin transparent oxide coating.

thermonuclear reactions: reactions that occur within atoms due to fusion, releasing an immensely concentrated amount of energy.

thermoplastic: a plastic that will soften, can repeatedly be moulded it into shape on heating and will set into the moulded shape as it cools.

thermoset: a plastic that will set into a moulded shape as it cools, but which cannot be made soft by reheating.

titration: a process of dripping one liquid into another in order to find out the amount needed to cause a neutral solution. An indicator is used to signal change.

toxic: poisonous enough to cause death.

translucent: almost transparent.

transmutation: the change of one element into another.

vapour: the gaseous form of a substance that is normally a liquid. For example, water vapour is the gaseous form of liquid water.

vein: a mineral deposit different from, and usually cutting across, the surrounding rocks. Most mineral and metal-bearing veins are deposits filling fractures. The veins were filled by hot, mineral-rich waters rising upwards from liquid volcanic magma. They are important sources of many metals, such as silver and gold, and also minerals such as gemstones. Veins are usually narrow, and were best suited to hand-mining. They are less exploited in the modern machine age.

viscous: slow moving, syrupy. A liquid that has a low viscosity is said to be mobile.

vitreous: glass-like.

volatile: readily forms a gas.

vulcanisation: forming cross-links between polymer chains to increase the strength of the whole polymer. Rubbers are vulcanised using sulphur when making tyres and other strong materials.

weak acid: an acid that has only partly dissociated (ionised) in water. Most organic acids are weak acids.

weather: a term used by Earth scientists and derived from "weathering", meaning to react with water and gases of the environment.

weathering: the slow natural processes that break down rocks and reduce them to small fragments either by mechanical or chemical means.

welding: fusing two pieces of metal together using heat.

X-rays: a form of very short wave radiation.

Index